YOUR

PUREBRED PUPPY

A Buyer's Guide

MICHELE LOWELL

Henry Holt and Company   New York

Henry Holt and Company, Inc.
Publishers since 1866
115 West 18th Street
New York, New York 10011

Henry Holt® is a registered trademark
of Henry Holt and Company, Inc.

Copyright 1990 by Michele Lowell
All rights reserved.
Published in Canada by Fitzhenry & Whiteside Ltd.,
195 Allstate Parkway, Markham, Ontario L3R 4T8.

Library of Congress Cataloging-in-Publication Data
Lowell, Michele.
Your purebred puppy : a buyer's guide / Michele Lowell.—1st ed.
p.   cm.
1. Dog breeds.   2. Dogs—Buying.   I. Title.
SF426.L69   1990   90-4349
636.7—dc20   CIP
ISBN 0-8050-1411-X
ISBN 0-8050-1892-1 (An Owl Book: pbk.)

Henry Holt books are available for special promotions
and premiums. For details contact: Director, Special Markets.

First published in hardcover in 1990 by
Henry Holt and Company, Inc.

First Owl Book Edition—1991

Designed by Claire N. Vaccaro

Printed in the United States of America
All first editions are printed on acid-free paper.∞

1   3   5   7   9   10   8   6   4   2
7   9   10   8   6
pbk.

To the three German Shepherds
who will always be dearest to my heart:

Kelly (Handreit's Kelly, C.D.X.),
Luke (Cochituate's Warrior Apollo, C.D.),
and Gretchen (Golden Eagle's Northern Light, C.D.).
Someday along God's garden path,
you'll hear me call your names.

And to my special friend, Mike:
Your thoughtful ideas and
gentle encouragement will always be there.

CONTENTS

STEP TWO: CHOOSING THE RIGHT BREED   15

3.   Evaluating Your Personality and Lifestyle   17

The importance of choosing the right breed · The pitfalls of choosing the wrong breed · The Questionnaire · If you want your dog to point birds, hunt rabbits, herd sheep · The truth about watchdogs

4.   Using the Breed Profiles   24

How to match your Questionnaire answers against the breed profiles · How to score each breed for suitability · What to do if you've had a bad experience with a breed · How to narrow your list of suitable breeds to three final contestants

5.   Choosing Your Breed   33

Making your final selection of the right breed for your family

THE BREED PROFILES   35

The Sporting Group · The Hound Group · The Working Group · The Terrier Group · The Toy Group · The Non-Sporting Group · The Herding Group · The Miscellaneous Group · The Rare Breeds

STEP THREE: CHOOSING THE RIGHT BREEDER   211

6.   Why Should You Buy from a Breeder?   213

Sources for purebred puppies · The meaning of registration papers and pedigrees · The history and importance of breed standards · Comparing good breeders and poor breeders · The real problems with pet shops · The frauds of puppy mills · Your neighbor's litter · Visiting the animal shelter · Breed rescue leagues

ACKNOWLEDGMENTS

I would like to thank the clubs, breeders, and owners who contributed breed information and photos—they and their beautiful dogs are credited individually at the end of this book.

I would also like to thank my husband, Ernie, my agent, Jane Gelfman, and my editor, Channa Taub—their enthusiasm sounded sweet to this novice writer.

And thank you, Randy, for watching over my shoulder while I worked on the computer; you were great company, and now you probably know more about canine breeds than any other five-year-old in the world.

STEP ONE

Deciding to Get a Purebred Dog

Should You Get a Dog?

So you think you want a dog.

Let's consider your decision some more. Let's talk about what it will mean to you, to your family, and to your present life-style. And let's be very sure you're making the right decision, because there's much more to owning a dog than you might have thought.

There's *money*. The purchase price will run from nothing (for a puppy from a friend's litter), to $25 (for a puppy from an animal shelter), to $250 to $500 (for a puppy from a breeder).

Should you ever pay more than that? In some cases, yes. If you're interested in a dog for showing or breeding, you may have to pay more. If you're interested in a breed with whelping difficulties, you may have to pay more to cover the breeder's cost of Caesarean sections. If you're interested in a tiny breed, you may have to pay more because these breeds produce only one to two puppies per litter and the breeder cannot

spread his profit over the more typical five to eight pups. Difficult whelpers and tiny breeds include the Bulldog, Chow Chow, Pug, Shih Tzu, Pekingese, Japanese Chin, Boston Terrier, English Toy Spaniel, Chihuahua, and a few others.

If you're interested in a breed not recognized by the American Kennel Club (AKC), you may again have to pay more. There are hundreds of foreign breeds that have only a few specimens in the United States. Breeders interested in establishing a foreign breed in this country must spend money to import good representatives and you'll often pay extra for one of their pups.

But for a nice eight-week-old puppy of most breeds profiled in this book—for a puppy not destined to be a show dog, stud dog, or brood bitch, but simply a cherished member of your family—you shouldn't have to pay more than $500.

Try to stand firm on that because, unfortunately, the breeders of certain breeds have decided that their pups should be priced in the $600 to $1,500 range. Although their breeds are no more valuable or "special" than other breeds, these breeders create the illusion that they are. Then they laugh all the way to the bank as the unsuspecting public falls all over itself trying to own a "superdog."

A current superdog is the appealingly wrinkled Chinese Shar-pei. Once a rare breed on the verge of extinction, the Shar-pei is now so popular that litter advertisements fill page after page of *Dog World* magazine, and the breed is almost ready for full AKC recognition. Yet breeders continue to tout Shar-peis as a rare, unusual breed. In reality, the Shar-pei is like any other breed: His combination of personality traits (clean in nature, extremely independent, prone to dog-fighting) is considered desirable by some people and undesirable by others. Just like any other breed.

If you pay astronomical prices, you encourage inexperienced breeders to jump onto the bandwagon of the latest superbreed. These breeders are trying to make a fast buck, and they will be quick to abandon the current superbreed when the next fad breed comes along. Did you know that the Cocker Spaniel and German Shepherd were once considered superbreeds?

Don't allow any breeder to dupe you into thinking that just because you're paying forty times more than what you'd pay at the animal shelter, you must be getting a dog *worth* forty times more. A high price tag does not automatically buy you a superior dog. Much more important than how much a dog costs is how carefully he was bred and raised, how healthy he is, and how compatible his physical traits, behavior, and personality are with your own needs, wants, and life-style.

No dog is perfect. All dogs have good points and bad points. Each breed is right for some people and wrong for others. The goal of this book is to find the dog that is right for *you*. And he doesn't have to cost a thousand dollars.

How about the opposite, then—a low price tag? Can you find a nice puppy at the Humane Society or in a neighborhood litter? Of course you can. Many people have found their beloved dogs in just these ways. In the section on choosing the right breeder (Step Three), we'll talk about sources for purebred puppies: good breeders, poor breeders, pet shops, Humane Societies, rescue leagues, your boss's sister's neighbor's litter. You'll learn that the best source is usually a good breeder, but if you can't afford the $250 to $500 prices of most good breeders, you'll have to check out riskier sources and your search will have to be extra careful. But take heart: Spending less money—so long as you're careful with those important factors of good background and good health and compatible temperament—won't buy you less love.

Unfortunately, the money outlay doesn't end when you hand over your check and scoop up your puppy.

Vet bills start piling up right away—$50 to $100 per year for a checkup, vaccinations, heartworm blood test and preventive medication, and flea powder. And ear infections and skin allergies always seem to pop up at the most inconvenient times.

Another potential expense is ear cropping. Great Danes, Boxers, and Dobermans are a few breeds whose naturally hanging ears are surgically altered to stand erect. The surgery will cost $75 to $200 unless the breeder has already had it done or unless you decide, for humane reasons, not to have it done at all. Ear cropping is cosmetic surgery; it does not *have* to be done. More about ear cropping, and a related subject, tail docking, later.

Once you get your puppy home from the vet, you'll have to start filling him with food. You can figure roughly upon weekly costs of five dollars for a small dog, ten dollars for a medium dog, and fifteen dollars for a large dog for a mixture of dry and canned food. (You'll also be sneaking a box of Bonzo Biscuits or Snoopy Snacks into the grocery cart, so better figure on an extra dollar or two.)

Other "doggy essentials" include a buckle collar with identification tag, a training collar, a six-foot leash of leather or nylon/cotton web, a brush and comb, nail clippers, dishes, Nylabones for chewing, a hard rubber ball, and *The Dog Owner's Home Veterinary Handbook*. Let's say $75 to $100 for the whole lot. A wire crate, $30 to $100 depending on size, is recommended as a housebreaking aid and den. Trimming shears and clippers, if needed for your breed, run about forty-five dollars, or you can pay a grooming shop twenty dollars every few months. If you take frequent business trips or if your family enjoys long vacations, boarding costs may be a factor at $5 to $10 per day. And the most important $40 to $50 you'll spend in your dog's life is enrollment in an obedience course when he's three to six months of age. So *money* is indeed involved in owning a dog.

Time and patience are involved, as well: time and patience for feeding your dog, for housebreaking him, for grooming him, for cleaning his yard, for walking him, for taking him to the vet, for playing with him, for training him, for talking to him, for loving him.

Don't try to deceive yourself about the amount of time and patience you need to live with a dog. It's hard to stand outside in the rain or snow every two hours trying to housebreak a puppy. It's harder still when he simply gazes around at all the sights, then wets on the floor as soon as you come back inside. It's hard to resist strangling him when he chews your shoes into sandals, but it's also hard to be firm when there he sits, sprawled awkwardly on one hip, cocking his head quizzically at you, panting and smiling.

It's hard to spend hours training him to sit and stay and come. It's harder still when you give him a sit command in front of your friends and he stares at you as though you had two heads. It's hard to clean the house for guests and then watch your dog track in mud, throw up on the carpet, splash water from his bowl, sweep his tail across the coffee table, and shed loose hairs all over the sofa. It's hard to bandage a bleeding paw or pull off a tick or use a rectal thermometer or race frantically to the vet in the middle of the night. And sometimes it's even hard to remember to rub his ears every night because he's waited all day for you to come home.

Time and patience. Be honest—do you lose your temper when the blockheads all around you don't seem to understand the perfectly clear things you're telling them? It's not a sin to have little time and less patience, but it *is* a hint that you should look for a different type of pet: a bird, a gerbil, fish.

If you believe that you have the time and patience, how about the *consideration*: consideration not only for the health and safety of your dog, but for the rights of other people? A considerate person trains his dog to be well mannered. A considerate person doesn't let his dog bark incessantly, threaten the mailman, chase the neighbor's cat, jump on visitors, or snatch ice cream cones from little children.

A considerate person doesn't let his dog run free, as in: "I'm going to let my dog run loose—he'll be happier." Is happiness being crushed by a car? Caught in a leg-hold trap? Chewed up in a dog fight? Blinded by stones pegged by bullies? Shot by the game warden for running deer or livestock? Poisoned by irate neighbors for digging up their lawn? Perhaps your happy dog will *cause* a tragic accident. Have you ever swerved sharply to avoid a dog dashing across the street? Was there a car coming in the opposite direction? Might there be next time? If you are so unconcerned about the lives of innocent people that you would allow your dog to risk causing injury or death, you should not have a dog. If you will not recognize the rights of other people in our crowded society, you should not have a dog. And if you will not protect the life of your dog by confining him safely, you should not have a dog.

And finally, the most important factor involved in owning a dog: *Love.* "Of course I'll love my dog," you say. And I'm sure you have the best intentions. But do you know *how* to love your dog? Will you know how to love him in ways that he will recognize and understand?

You and he are two different species. Since *you* are the more intelligent, you must love your dog and communicate with him in ways that he will understand. You must speak his language, because he can't speak yours. Every time you do something with your dog, every time you speak to him, you must ask yourself what his canine eyes are seeing and what his canine mind is understanding. Your interpretation of your actions is seldom the same as his interpretation, and his

interpretation is what counts. You must communicate so that *he* understands.

How do you show your dog that you love him? First and foremost, by recognizing him for what he is. God created your dog as a wonderful and unique creature with talents and abilities all his own. It is not only incorrect but wasteful and demeaning to consider him a furry person. His thought processes are very different from yours; he is much less complex and thus he seldom needs to read self-help books. He has basic instincts and basic ways of responding to actions. He has definite limits to his physical and mental capacities that you need to understand, first, so that you do not ask too much of him, and second, so that you do not ask too little. Before you can truly love your dog, you have to understand him.

Let's take a practical, close-up look at your dog. What does the world look like to him? Looking through his eyes is like looking through the eyes of a nearsighted person without glasses: The world looks fuzzy and vague. If you crouch down to his eye level, you'll get an idea of his perspective on the world: Things look towering, don't you think?

Can he see details? Although his vision is fuzzy, he may wag his tail when you smile and drop his tail when you frown, because most dogs are very observant at picking out details in your expression. You may think he is slinking away from your shredded slipper because he feels guilty, but usually he's reacting to your outraged expression and threatening body posture. He knows from past experience, not from any sense of guilt, that this particular facial expression and body posture don't bode well for him.

His peripheral, sudden-movement vision is far better than yours. Even when he's looking in another direction, he'll suddenly spy the wriggle of a mouse in deep grass. His night vision is better than yours because he can dilate his pupils more, and widening his pupils lets in whatever little light can be found. Can he see color or just shades of gray? The debate rages, but interestingly, many dogs seem to prefer rubber balls that are red, and many dogs are suspicious of black clothing.

What does the world smell like to him? Just as your world is full of sights, his world is full of smells. He relies upon his nose as you rely upon your eyes. He can pick out, from a dozen articles, the single one touched by his owner—and thus bearing his owner's scent—just as easily as you can pick out a green ball from a pile of blue ones.

He actually creates mental pictures based on smells. When the family car pulls into a campground where he has been before, he will get excited. He can't read the campground sign, but he recognizes the tree and wildlife scents, and from past experience he associates these with a fun-filled weekend: Oh, boy! But when you lead him toward the veterinarian's front door, he will balk. He can't read the veterinary sign, but he recognizes the medicinal scents, and from past experience he associates these with examining tables and probing fingers: Oh, no! (He also recognizes and reacts to your happy "campground" attitude and your nervous "veterinary" attitude.)

What does the world sound like to him? He hears much better than you do—softer sounds over greater distances at higher frequencies. Some dogs burrow under the blankets long before you hear the faint roll of thunder. Some dogs hear imminent avalanches and earthquakes.

Use a low, pleasant voice when you talk to your dog. Just as with children, who must of necessity stop their crying in order to hear what you're saying when you speak in a quiet voice, dogs tend to pay closer attention. Does he understand your words? Not unless you link them to the correct object or action. Puppies learn language just as babies do: by your saying "apple" over and over again while holding up an apple. Until you connect a word with the correct object or action, words are just meaningless sounds. Imagine yourself listening to a conversation in a foreign language, where the words are not connected to anything concrete.

How do dogs communicate with one another and with their human families? Mostly they use body postures and facial expressions; sadly, their human families usually don't recognize what the dog is trying to say. If your dog raises his ears and tail and hackles, stiffens his body, and takes slow, deliberate, stiff-legged steps, he is indicating aggressiveness, saying, "I'm in charge." If he crouches low, wags his tail hesitantly between his legs, flattens his ears, draws his lips into an ingratiating smile, lifts his paw beseechingly, rolls onto his back, perhaps even dribbles urine, he is indicating submissiveness, saying, "Whatever you want is okay with me." If he raises his hackles and lifts his lip, but also crouches and tucks his tail, he feels both threatened and afraid. His survival instinct is telling him, "Attack! Flee! Bite! Run!" and thus his mixed body signals. You must recognize these body postures if you are to understand your dog.

You must also recognize his most important instinct, the pack instinct. It is so powerful and so instrumental in making your dog a full-fledged member of your family that you need to fully understand it in order to live comfortably with your dog.

Dogs are sociable animals who like to live with other sociable animals in a group or pack. Within the pack there is always a hierarchy. At the top is the most dominant animal, the pack leader, who establishes and enforces rules, carries out discipline, and makes decisions. Next in line is the number-two animal, then the number-three animal, right down to the most submissive one of all. Pack animals are not unhappy with this hierarchy—on the contrary. Knowing exactly where they stand with one another and exactly what the rules are makes for a strong feeling of security and fellowship. The survival of the pack depends upon every member being able to handle his respective position.

The pack instinct is the main reason dogs wedge themselves so completely into our families, rather than prowl along the fringes, like most cats. Cats are solitary animals who like to do their own thing. Dogs are pack animals who like to belong. That one instinct makes a tremendous difference in the way each pet—dog or cat—should be raised. It also creates a greater responsibility for you than if you'd decided to get a cat.

How so? When a dog joins your family, even if your family is only yourself, a pack is formed. Oh yes, in his mind it certainly is, and his instincts compel him to seek out its structure. Who is the leader? Who is the follower? If you don't establish yourself as the leader, he may take the position himself, for most dogs are not comfortable in a leaderless world. Any good obedience instructor will tell you that the major cause of problems between owners and dogs is uncertain leadership.

Another major cause of problems is when there is *no* uncertainty about leadership because the *dog* has assumed the dominant role. This is an unpleasant state of affairs that can actually jeopardize your dog's safety, because there will be many times when you'll have to do things with him that he may not enjoy, like examine his teeth. You must be dominant enough so that he will allow you to do anything with him. He must accept *your* judgment as to what is best for him.

I can hear somebody out there wailing, "My dog would never let me touch his teeth!" Stop that. What are you going to say when he chokes to death because he wouldn't "let" you pull a stuck bone out of his throat?

And how are you going to train him? "Oh, but he loves me!" you protest. "Dogs want to please the people they love!" Don't all obedience instructors wish! Dogs want to please only the people they respect: leaders. Dogs will coexist with, ignore, or challenge followers. They will love you either way, for they do not in any way equate love with respect. They love blindly; they respect only those who have earned it. So teaching them to respect you will in no way diminish their love for you, and teaching them to respect you is mandatory if you are to take proper care of them.

Do you wish you'd decided to get a cat?

A complication: In a multiowner household, even if one person establishes himself as the leader, the dog may test another person. He may respect the person he recognizes as being above himself on the ladder, but not the person he places below himself. Obedience instructors often hear: "Duke listens to me but not to my wife." Or vice versa. Dogs always seem to pester guests who don't like them because they pick up the guest's uncertainty and test to see if they can dominate. Sometimes it's done playfully, sometimes seriously. Remember, in a dog's mind, life is filled with leaders and followers, and he likes to find out who's who.

A second complication: children. Dogs usually consider children below themselves in pack position. A child should never try to boss a dog who considers that child a subordinate, because the dog may defend against what he sees as a threat to his pack position. A child should *gradually* be moved up the ladder above the dog by feeding the dog, holding the leash, and enforcing simple commands. Young children should never wrestle with a dog or play tug-of-war or chase. These games only reinforce the dog's speed and strength compared to the child's.

A third complication: In a multidog household, if an unruly dog dominates an obedient dog, the obedient dog may follow the unruly one into bad behavior. It's important that both dogs see you as a strong authority figure so that they don't ignore you to follow each other.

How does one establish oneself as a pack leader? Fortunately, with most dogs it's very easy. Most dogs are happy to be followers if you are willing to be the leader. You demonstrate your leadership with a series of little things that may not seem like much to you, but that your dog definitely sees as connected to a pack leader:

- Teach him obedience exercises.
- Frequently make eye contact with him.
- Don't play aggressive games like wrestling or tug-of-war.
- Speak in a low, pleasant, firm voice.
- Use short, simple words and equate them directly to specific, concrete objects or actions.
- Be utterly consistent with your house rules (if he can't climb on the couch on Monday, he can't climb on it on Tuesday).
- Praise him when he does something you like (even if that something is merely lying quietly at your feet).
- Stop him *immediately* when he does something you don't like.
- Insist that he accept being groomed.
- Teach him to take food gently from your hand—no grabbing.
- Insist that he behave in public.

I can hear you breathing a sigh of relief at how easy those tasks sound. You were probably afraid, with all that pack-leader talk, that you'd have to abuse your dog to convince him who was boss. Not at all. Psychological superiority is far more important than physical superiority. Convince the twenty-pound puppy with a firm voice and confident attitude that you are in charge, that you will make all the decisions, that you have everything under control, and the hundred-pound adult will never question you.

You also demonstrate your leadership by immediately putting a stop to any sign of canine dominance. Such dominance means your dog is testing your leadership. This is apt to occur more in some breeds than in others, and more in some dogs than in others. Not all dogs want to dominate, and you should be gentle with dogs who already roll over onto their backs or look apologetic at any sign of displeasure from you. They have accepted your leadership, and any further display of strength on your part would be unfair according to their social code.

What does a dominance test look like? It sounds ominous, but don't worry—your dog is not going to pounce on you in the middle of the night. Just as your own buildup of leadership is a series of little things, your dog's buildup of leadership is too. He may protest giving up a toy. He may shove his head at you, insisting that you pet him. He may utter his displeasure if you walk by his dish or bump into him. He may mouth your hand when you try to clip his nails. He may bark at you when you give a command. He may lunge at strangers, utterly ignoring your attempts to stop him. Urinating on your belongings is an unmistakable dominance challenge!

How should you respond? Sharply. Just as a child learns that when he puts his hand close to the fire he experiences a sharp discomfort that suggests, "It would not be to your advantage to do that again," so your dog should learn that when he attempts to dominate you, he experiences a sharp discomfort that does not merely suggest, but *commands*, "It would not be to your advantage to do that again."

Pack leaders discipline calmly. They don't nag, shout, point fingers, wave rolled-up newspapers, or let the discipline drag on. They do it and are done with it and they bear no grudges. In a sharp, guttural voice, say "NO! STOP!" Shake his collar, jerk his leash, or clip him sharply under the chin with your fingers. With a young dog, an effective disciplinary method is to grasp the loose skin on either side of his neck near his face, lift his front end off the ground, maintain eye contact until he looks away, then release him.

Discipline cannot leave the slightest doubt in the dog's mind that you will not accept this behavior. A dog will always weigh his enjoyment of a behavior against the discomfort of your correction, and he will decide which takes precedence. If you are to understand your dog, communicate with him, and love him, you need to recognize that he learns almost everything from cause and effect, from event and result, from your reaction to his action.

We've talked about *money, time and patience, consideration,* and *love.* Those are the responsibilities of dog ownership. Those are the things that you as an owner must give to your dog. And what do you get in return?

The rewards can't be measured tangibly, but that's the case with so many rewards of great value. A dog offers you pride in his appearance and behavior. A dog offers you the opportunity for accomplishment: that of having raised and trained him well. A dog offers you fun and play. An old saying goes, "With a dog, not only can you make a fool of yourself without him laughing at you, but he will make a fool of himself, too."

A dog offers you companionship and devotion and a unique, unconditional love that is neither greater nor lesser, but simply different, from its human counterpart. An epitaph on a beloved pet's gravestone reads: "The reason I loved him is plain to see; with all my faults, he found beauty in me."

Senator Graham Vest said it best when he spoke his famous *Tribute to the Dog* over a century ago:

Gentlemen of the Jury: The one absolutely unselfish friend that man can have in this selfish world, the one that never deserts him, the one that never proves ungrateful or treacherous, is his dog. A man's dog stands by him in prosperity and in poverty, in health and in sickness. He will sleep on the cold ground, where the wintry winds blow and the snow drives fiercely, if only he may be near his master's side. He will kiss the hand that has no food to offer, he will lick the wounds and sores that come in encounters with the roughness of the world. He guards the sleep of his pauper master as if he were a prince. When all other friends desert, he remains. When riches take wings and reputation falls to pieces, he is as constant in his love as the sun in its journey through the heavens.

Is this trade-off between what you give and what you get worth it? That is what you must decide before you buy a dog. Those who decide no are undoubtedly missing out on a lot, but there are many legitimate reasons not to get a dog.

You may not be able to afford one, although so long as food and health needs are unfailingly taken care of, other costs can be kept down with a little thought. Don't deny yourself the love and companionship of a dog simply because of limited funds.

You may work long hours or participate in so many activities that little time would be left for the dog.

You may travel and cannot bring a dog along or board him.

You may like an immaculate environment.

You may consider a dog to be a possession that can be locked in the garage or tied outside when you're not showing him off.

In all these cases, you should *not* get a dog. Both of you will be unhappy.

And don't let a friend lay on you that classic pressure: "Don't you want to teach your children responsibility?" If you yourself are not committed to caring properly for a dog, what you will be teaching your children is irresponsibility. If feeding and exercising and training are neglected, if the dog is bred indiscriminately and new, unwanted lives carelessly brought into the world, if he is dropped off at the pound when he becomes inconvenient, you will be teaching your children that the life of an animal is not worth much, that it can be used and tossed aside at whim. Many children progress from a callous and uncaring attitude toward the life of an animal to a callous and uncaring attitude toward all life.

So, yes, you have a perfect right to decide no.

Those who decide yes should be willing and able to take on the pleasures *and* the responsibilities, and willing and able to make a commitment.

That's exactly what this is—a commitment to be fully responsible for the life of another living creature. If you have a child, you may already understand that this type of commitment requires that you love, provide, teach, share, enjoy, discipline, worry, laugh, cry, and eventually let go.

But there is a difference—a big one. With a child, your relationship changes as he grows; as he gradually becomes independent, you gradually let go. Your dog, on the other hand, regards you as his whole world throughout his short life, and the letting go is sometimes sudden and always final. Someday you may find yourself standing in the veterinarian's office for the last time, having reached the final and most desperately difficult stage of the commitment you made when you brought home that little puppy.

So you think you want a dog. Do you know for sure now?

If you've weighed the advantages and disadvantages, the pluses and minuses, the responsibilities and rewards, and come out on the yes side, we have to find you a dog: the right dog for you.

Should You Get a Purebred

or a Mixed Breed?

Purebred or mixed: Which is the better dog? Each side has staunch proponents, but in the end, every dog is a combination of heredity, environment, and a plain old roll of the dice. If two dogs stand side by side, one purebred and one mixed, either may be the more beautiful, the more intelligent, the better family pet. When one purebred is compared to one mixed breed, either may come out ahead. It is only as a group that purebreds have a big advantage over mixed breeds. And the advantage is that of predictability.

People began selective breeding based strictly on working ability. The best herding dogs were bred to one another, and the puppies grew to be fine herders. The best hunters were bred to one another and the puppies were found to have strong hunting instincts and abilities. Only later were dogs selected for specific physical features—prick ears in one herding breed, hanging ears in another—and specific temperaments— friendliness in one hunting breed, aloofness in

another. Selective breeding, in its simplest terms, means breeding *like* to *like* to get more little likes!

But in more accurate terms, selective breeding means breeding *like genes* to *like genes*. Oh-oh, not that old high school science of genetics! Just five quick points:

1. Traits such as herding and hunting instinct, prick and hanging ears, friendliness and aloofness, are contained in genes, which pass from parent to offspring during conception.
2. So when dogs with a certain trait were bred, it was expected that the genes carrying those traits would be passed along to the puppies. Sounds logical, doesn't it?
3. It didn't always work. Sometimes two dogs with the desired trait—say, prick ears— couldn't seem to pass prick ears along to their offspring. How could that be? The two dogs, Mom and Dad, had the prick ears,

right? That meant they had to have the "prick ears gene," right? And that gene had to pass along to their offspring, right?

4. Not necessarily! Let's look at Dad first. Since Dad had his own Mom and Dad, he had to have received *two* genes for ears: one from his Mom and one from his Dad. Since he'd turned out with prick ears, *one* of those genes had obviously been for prick ears, but the other one might have been for hanging ears. So he could have passed either "prick" or "hanging" along to his own offspring. The same with Mom. Remember dominant and recessive genes from high school biology? Two brown-eyed parents can have a blue-eyed child if both parents have a recessive gene for blue eyes and these recessive genes come together in the child.

    So the traits that a puppy inherits depends upon the gene *combinations* he inherits, and since those don't show on the outside, a lot of trial breeding had to take place until the game of whose-genes-meshed-properly-with-whose could be figured out.

5. When dogs with the right gene combinations were finally bred together, whole litters began sharing not only a set of the same *traits*, but a set of the same *genes*. With no more undesirable genes in the picture, the desirable genes became *fixed* and could be passed on predictably to the next generation, and the next and the next, so that the breed could continue to look and act and work as its developers wanted.

    For example, Daisy is a Golden Retriever, a large dog with a long golden coat, hanging ears, and a wagging tail. Golden Retrievers have fixed genes for large size, a long golden coat, hanging ears, and friendly personality. When Daisy is bred to Toby, another Golden, their puppies will inherit their fixed genes and turn out large and friendly, with a long golden coat and hanging ears. You have reproduced Golden Retrievers.

This is the science behind all purebreds: fixed, predictable genes. And this is the advantage of a purebred puppy: Because he is born with a fixed set of genes, we know what he will look like and act like as an adult.

A mixed-breed puppy, on the other hand, inherits random, unfixed genes from his parents. Whether these genes contain traits which are going to appeal to you cannot be determined until the puppy is grown. And since these genes are unfixed and often wildly contrasting, they cannot be passed on predictably to his own offspring. Thus, a mixed breed should never be bred, for no matter how beautiful or intelligent he is, he does not have the fixed genes to reproduce his appearance or temperament.

Are there any advantages to a mixed breed over a purebred? Yes. Because his genes are mixed, he doesn't have as many medical disorders given to him via heredity. He often has a more middle-of-the-road temperament, while many purebred temperaments are *very* friendly, *very* independent, *very* energetic, and so on.

Mixed breeds make wonderful pets—no doubt about it. I have owned them all my life and love them dearly. The trick is in trying to choose, from the myriad gene combinations out there, the mixed breed that is going to turn out to fit your family. That's where the smaller and more predictable group of purebreds—who also make wonderful pets—comes into its own.

Simon & Schuster's Guide to Dogs lists over three hundred breeds, but many of those are foreign breeds not present or not common in the United States. So you will be choosing, first, from 153 breeds and varieties recognized by the American Kennel Club (which is not a club at all, but the largest purebred registry in the United States). It's not that the AKC feels that the remaining breeds in the Simon & Schuster *Guide* are mixed breeds! The ones they don't "recognize" simply don't have enough (if any) representatives in this country to warrant the trouble and expense of registering them, offering classes for them at dog shows, and so on. They're still purebred, however.

The United Kennel Club (UKC), the second-largest registry in this country, registers many of the same breeds as does the AKC, plus a few more. So, to our list of AKC breeds you'll be choosing from, we've added some UKC breeds—the ones most likely to suit most families. We've also added some nice breeds that for various reasons have not yet been recognized by either the AKC or the UKC. But they *are* purebred. As you know now, purebred simply means that a type of dog has been so consistently bred to other members of the same type that the genes have been fixed and will be passed predictably from parent to offspring.

Many of those predictable genes concern temperament and behavior. Our next step, then, is to find out what kind of temperament and behavior you're looking for in a dog so that we can find the one purebred, out of the 175 profiled in this book, that has the genes and characteristics that will best match you.

STEP TWO

Choosing the Right Breed

Evaluating Your Personality

and Lifestyle

You must understand the importance of choosing the right breed, because a major cause of problems between an owner and his dog is that the dog doesn't suit the owner and/or the owner doesn't suit the dog. Some examples of incompatibility are an easygoing man choosing a strong-willled Chow; a woman who hates to groom choosing a shaggy Old English Sheepdog; a sociable family choosing a suspicious Chihuahua.

Some people choose a breed because they like its muscular appearance or beautiful coat or cute face, or because there's a litter handy, or because "My uncle had one who could practically talk." But in a week or a month or a year, these people may well discover, to their dismay, that they've started a ten- to fifteen-year commitment with an incompatible breed. Owner and Dog will end up in a discouraging battle as Owner struggles to change Dog's natural characteristics to ones that he would like better.

"He digs in my flower beds!" wails the gardening enthusiast. Of course he does—he's a Fox Terrier, bred to dig vermin out of their underground dens. It's in his genes.

"He barks and barks!" wails the writer who can't concentrate on her work. Of course he does—he's a Norwegian Elkhound, bred to bark continuously when cornering moose. It's in his genes.

"He's so aloof with my friends!" wails the society woman. Of course he is—he's a Borzoi, bred not as a sociable house pet but as a tough, independent hunter of wolves. It's in his genes.

Remember that most breeds were developed for a reason, and the reason usually had to do with working ability. The instincts and characteristics that best suited that working ability are still going to come through today because of the purebred's fixed, predictable genes.

You might curse those fixed genes as you're refilling your flower bed or plugging cotton in your ears or apologizing to your miffed friends, but these three owners could just as easily have made those fixed genes work for them instead of against them. The gardener could have chosen a dignified Pekingese. The writer could have chosen a quiet Mastiff. The society woman could have chosen a friendly Bichon Frise. They just didn't know.

But you do. You can start your owner-dog relationship off on the right foot by evaluating yourself and your lifestyle, deciding which canine characteristics would best suit you, and then choosing a breed that has those characteristics. Of course, almost any temperament and behavior can be modified and controlled with proper handling and training, but it will be less frustrating all around if you start by choosing a breed who is at least reasonably compatible with your wants and needs.

The questionnaire that begins on this page asks about your needs, wants, personality, and lifestyle. Your answer to each question will reveal a specific trait you're looking for in a dog. As you answer each of the ten questions, you'll be instructed to write a word or a couple of words on a sheet of paper. So start by taking out a sheet of paper and writing at the top, "Traits I'm looking for in a dog." List the following categories down the left side of the page:

1. Experience Required:
2. With Children:
3. Size:
4. Coat:
5. Exercise Required:
6. Trimming/Clipping Required:
7. Amount of Shedding:
8. Activity Indoors:
9. Ease of Training:
10. Sociability with Strangers:

When you've answered all ten questions, you'll have a list of ten traits you're looking for in a dog.

THE QUESTIONNAIRE

1. EXPERIENCE REQUIRED

Answer T(rue) or F(alse).
— I have had experience and success in training a dog to be well mannered and obedient.
— I found the discussion on pack leadership interesting, and I understand how to be a leader with a dog.
— I think most parents are too permissive or lackadaisical.
— I like to make decisions.
— I often feel I have better judgment than others.
— I would make a good leader in group projects.
— I hold opinions on most subjects.
— I will stand up for myself when I think I'm right.

If you answered True to seven or eight of these statements, you would probably be equipped to handle a breed who tried to test your leadership. Write the following two phrases beside Category 1: *For experienced owners only* and *Fine for novice owners.*

Otherwise, write *Fine for novice owners*— you'll still have the majority of the breeds to choose from.

2. WITH CHILDREN

Do you have children or oft-visiting relatives under age eight? Are there neighborhood children under age eight who would regularly come into close contact with your dog? Are you planning to have a baby within your dog's lifetime? If you answer yes to any of these questions, write *Good with children* and *Good with children if raised with children* beside Category 2.

If you answer no to all of these, write *Good with children* and *Good with children if raised with children* and *Good with older, considerate children.*

Caution: All children, young or old, must be taught to be gentle with all dogs. Buying a tolerant, easygoing breed does not give your toddlers license to be rough and abusive. Even breeds considered good with children *can bite* if they are hurt, teased, or frightened.

3. SIZE

Many people have preconceived notions about various sizes of dogs. They may feel that all little dogs are cuddly, or the reverse—that all little dogs are snappy. They may feel that all big dogs are protective, or the reverse—that all big dogs are easygoing. Not true. Almost every type of temperament is available in almost every size dog. However, some generalities should be considered:

Little dogs, about the size of a Toy Poodle, can live comfortably in tight quarters and do not need much (if any) outdoor exercise, although most would not turn down a daily walk. They do not always need obedience training. They are seldom objected to by neighbors and are often allowed by landlords. They are easy to clean up after and easy to travel with.

Small dogs, about the size of a Cocker Spaniel, can live comfortably in tight quarters but need a small yard or a short walk every day. They need some obedience training. They are seldom objected to by neighbors and are often allowed by landlords. They are easy to clean up after and easy to travel with.

Medium dogs, about the size of an English

or Welsh Springer Spaniel, need an outdoor yard or a good walk every day. They need obedience training and a chance to run (in a safe, enclosed area) now and then. They are seldom objected to by neighbors and are often allowed by landlords. They require a little work to clean up after. They can travel with you most places.

Large dogs, about the size of a Golden Retriever, need an outdoor yard or a long walk every day. They need obedience training and a chance to run (in a safe, enclosed area) now and then. They are sometimes objected to by neighbors and are often not allowed by landlords. They require a lot of work to clean up after. They can be difficult to travel with.

Giant dogs, about the size of a Saint Bernard or Great Dane, need an outdoor yard or a long walk every day. They need obedience training and a chance to run (in a safe, enclosed area) now and then. They are very often objected to by neighbors and are seldom allowed by landlords. They require much work to clean up after. They are very difficult to travel with. Also, larger dogs are more expensive to feed and board.

Which sizes would be acceptable to you? Write your choices (as many as you like) beside Category 3.

4. COAT

Dog coats vary in length and texture.

Smooth (Doberman, Beagle, Chihuahua): needs only a quick brushing once a week.
Short (Labrador Retriever, Pembroke, Welsh Corgi): needs five to ten minutes brushing once a week.
Thick medium in length (German Shepherd, Siberian Husky, Pomeranian): needs ten to fifteen minutes brushing twice a week.

Feathered medium in length (Golden Retriever, Cocker Spaniel): needs ten to twenty minutes brushing and combing twice a week.

Long (Afghan, Collie, Lhasa Apso): needs ten to twenty minutes brushing and combing every other day.

Wiry with bushy eyebrows and a beard (Schnauzer, Cairn Terrier): needs ten to twenty minutes brushing and combing twice a week.

Curly (Poodle, American or Irish Water Spaniel, Bichon Frise): needs ten to twenty minutes brushing and combing twice a week.

Which coats would be acceptable to you? Write your choices beside Category 4.

5. EXERCISE REQUIRED

If you have no outdoor yard and don't plan on taking your dog for many regular walks, write *Low* beside Category 5. If you have an outdoor yard or will take your dog for regular walks and occasional runs, write *Low* and *Medium*. If you have an outdoor yard and will regularly take your dog jogging or hiking or biking, write *Low* and *Medium* and *High*. Almost all breeds enjoy hard exercise and athletic activities; the exceptions are short-nosed bulldog-type breeds and large, heavy-coated breeds, who may suffer respiratory and heatstroke problems from too much exertion in warm weather.

6. TRIMMING/CLIPPING REQUIRED

Which one of these statements fits you best?

a. I only want a breed with a no-fuss coat (no trimming).

b. I wouldn't mind a breed that needed some minor trimming of straggly hairs every three to six months, but I wouldn't want one that needed electric clipping every two to three months.

c. I wouldn't mind a breed that needed clipping every two to three months. I'd just learn to use clippers or I'd pay a groomer.

d. I'd definitely prefer a stylishly clipped dog.

If you chose a, write *Low* beside Category 6. If you chose b, write *Low* and *Medium*. If you chose c, write *Low* and *Medium* and *High*. If you chose d, write *High*.

7. AMOUNT OF SHEDDING

How much vacuuming of dog hair are you willing to do? If none, or if you hate the idea of hair on your clothes and furniture, or if anyone in your family is allergic to dogs, you need a breed that sheds little, so write *Low* beside Category 7. *Note:* Allergic people are often helped by rubbing Allerpet on their dog—see your pet-supply shop.

If vacuuming a few times a week during shedding season is okay and if you wouldn't die of embarrassment if a visiting friend left with a couple of hairs on her pants, write *Low* and *Medium*.

If you wouldn't mind vacuuming every day during shedding season and pulling out handfuls of hair from your dog's coat, write *Low* and *Medium* and *High*; in other words, you would accept any amount of shedding.

8. ACTIVITY INDOORS

Answer T(rue) or F(alse).

— I tend to be restless or fidgety when I'm not busy.

— I've been told I have nervous energy or a high metabolism.
— I don't like to sit still for very long.
— I use a lot of gestures when I talk.
— I am a fast talker.
— I am a fast walker.
— I am a fast eater.
— I get tense or upset when I'm stuck in traffic.
— I'm annoyed by active, inquisitive children.

If you answered True to seven to nine of these statements, an active dog would drive you crazy. Write *Low* beside Category 8.

If you answered True to four to six of these statements, write *Low* and *Medium*.

If you answered True to zero to three of these statements, you could probably live with any degree of activity, so make your decision based on the following: If you'd prefer a dog who is lively and inquisitive in the house, who likes to play with you and follow you around, write *High* and *Medium* beside Category 8. If you'd prefer a dog who is not overly lively, write *Low* and *Medium*. If it really doesn't matter to you, write *Low* and *Medium* and *High*; in other words, any degree of activity is fine.

9. EASE OF TRAINING

Which one of these statements fits you best?

a. I'm unwilling or unable to do any obedience training with a dog.
b. I'm willing to train a smart and willing-to-please dog, but I'd become discouraged with a slow learner or a mildly stubborn dog.
c. I'm willing to train a smart and willing-to-please dog, a slow learner, or a mildly stubborn dog, but I wouldn't feel confident working with a dog who was very self-willed, independent, or inattentive.
d. I'm confident that I can learn to train any

dog. I'll find a skilled instructor or an excellent book, and I won't get impatient or discouraged even if he is difficult to train.

If you chose a, check back to Category 3 (Size) and be sure it says *Little*, because only little dogs can do without any training. Then write *High* and *Medium* beside Category 9, because you'll need a dog who is reasonably responsive all on his own.

If you chose b, write *High*. If you chose c, write *High* and *Medium*. If you chose d, write *High* and *Medium* and *Low*.

Any breed can be trained successfully to the highest level of obedience, and in experienced hands most breeds are reasonably easy to work with. However, with a novice trainer at the other end of the leash, some breeds have proven to be easier than others to work with and less inclined to take advantage of mistakes made by their inexperienced owners.

10. SOCIABILITY WITH STRANGERS

How would you like your dog to behave with people? If you'd like him to be friendly, to trust people, to bark when someone comes to the door but then welcome them inside, write *High* beside Category 10.

If you'd like him to be reserved or aloof, to approach people in his own time, to observe them from a short distance, write *Low*.

If either friendly or reserved is fine, write *High* and *Medium* and *Low*.

Ten questions answered! You now have a list of ten traits to look for in a dog.

But before we start combing through all 175 breeds and varieties to find the ones that match your list, let's make sure you *can* consider all 175.

There are two special circumstances that would require you to eliminate the great majority of the breeds before you start comparing.

Special Circumstance #1

Are you planning to use your dog to hunt birds or small game, to herd or guard livestock, or to compete in the German protection sport of Schutzhund? If no, move on to Special Circumstance #2. If yes, you won't have all 175 breeds and varieties to choose from, because only certain breeds are capable of performing such specific, demanding work. People who are already involved in these activities are your best advisors to those breeds, so you should put your search on hold until you have had a chance to talk to those people.

How do you find them? You can start by calling the AKC in New York City and asking for information on hunting trials and herding trials. For information on Schutzhund, pick up *Dog World* magazine, look for the display ad for *Dog Sports* magazine, and order a sample copy. Call a veterinarian, animal shelter, or boarding kennel and ask for the name and number of your local AKC breed or obedience club. You're searching for information about your activity, books to read, or further contacts. Follow all leads, and keep asking, "Do you know anybody who might know anything about . . . ?"

Once you know something about the activity and about the breeds best able to perform it, you can continue your search to find out which breeds of those that are capable of performing the activity would be the best match for you.

Generally, breeds used for bird hunting include pointers, retrievers, setters, and spaniels. Breeds used for game hunting include the Basset Hound, Beagle, Black and Tan Coonhound, American and English Foxhounds, Harrier, and some terriers. Breeds used for herding include the Australian Shepherd, Border Collie, English Shepherd, Norwegian Buhund, Petit Basset Griffon Vendeen, Polish Owczarek Nizinny, Swedish Vallhund, and breeds in the AKC Herding Group. Breeds used for guarding livestock include the Great Pyrenees, Komondor, Kuvasz, Louisiana Catahoula Leopard Dog, and some breeds in the AKC Herding Group. Breeds used in Schutzhund include the Belgian herding breeds, Bouvier des Flandres, Boxer, Doberman Pinscher, German Shepherd, Giant Schnauzer, and Rottweiler.

Special Circumstance #2

Do you want your dog to be a watchdog? First, let's define "watchdog." A watchdog barks when he hears someone or something near his property. Because every breed can and usually will bark at that time, every breed is a potential watchdog. (The exception is the Basenji, who cannot bark.) If an individual dog happens not to bark, that is usually the individual, not the breed.

The vast majority of burglars will shy away from homes in which any dog is barking; thus, barking—even the barking of a little dog—is all the "watching" that most families need. If you're looking for a dog whose appearance and behavior would give burglars even greater incentive to stay away, fine. Choose a medium or large or giant breed who is reserved rather than friendly with strangers.

If you're looking for a dog who will do more than bark and look imposing, please consider the following very carefully: Some breeds are naturally protective, but they must still be taught *how* to protect and *where* and *when* to bite. Professional training is expensive, and the result is the equivalent of a loaded gun. Guns often do more damage to the innocent than to the guilty, and a dog has no safety catch and cannot be unloaded. (Proponents of guard dogs argue that a dog, unlike a bullet, can be recalled, but they will also readily agree that few dogs are that well trained.) And training carries no guarantee of protection because a professional criminal has many ways of dealing with a dog: Mace or a bullet, for example.

Perhaps you want a naturally protective breed but don't want to go to the trouble and expense of professional training. Perhaps you believe that an untrained Doberman will protect you just as well as a trained one. You may be in for a shock. Let's say the burglar isn't even a professional with Mace or a gun. Let's say he's simply got enough guts to ignore your barking Doberman and break into your house. He's promptly bitten. Don't you agree that a man bold enough to enter a home in which a big Doberman is barking is going to fight back?

You bet he is. And since an untrained dog usually bites at the leg, the burglar, with both hands free, will knock your beloved dog's brains out with the nearest chair or stab him with a pocket knife. A few really tough dogs might continue to make a fight of it, but most inexperienced dogs who suddenly find themselves in a real fight with a human being will back down. Now the tables have turned as the angry burglar chases your bewildered dog with murderous intent.

Many a protective dog has lost his life trying to defend his owner's TV set, while the dog who is "so friendly he'd lead the burglar to the silverware" or "so shy he'd hide under the bed" escapes unharmed. Thus it may actually be a disadvantage to own a naturally protective breed, for no TV set is worth serious injury to a dog who is a member of your family.

But what if it's not your home and property that is endangered, but yourself by, say, a mugger on the street? Would you want your dog to protect you then? He may not have to, because most muggers, like most burglars, will shy away from anyone with a dog. Dogs can bark, and barking draws unwanted attention. But if you believe

firmly in self-defense and you want your protective dog to be similarly equipped, consider joining a Schutzhund club. Schutzhund is a rigorous German sport that combines tracking, obedience, and protection exercises. The training is exciting, and the result is a dog-owner team that works together in a close relationship.

Final recommendation: If all you want is a dog that barks, you can make your choice from among all the breeds. But if you must have a breed with some size, muscle, and a protective reputation, choose from: German Wirehaired Pointer, Chesapeake Bay Retriever, Irish Water Spaniel, Weimaraner, Norwegian Elkhound, Rhodesian Ridgeback, Akita, Boxer, Bullmastiff, Doberman, Giant Schnauzer, Great Dane, Great Pyrenees, Komondor, Kuvasz, Mastiff, Rottweiler, Standard Schnauzer, Airedale Terrier, American Staffordshire Terrier, Irish Terrier, Kerry Blue Terrier, Staffordshire Bull Terrier, Chow Chow, Australian Cattle Dog, the Belgian herding breeds, Bouvier des Flandres, Briard, German Shepherd, Puli, Australian Shepherd, Chinese Shar-pei, Chinook, Leonberger, and Louisiana Catahoula Leopard Dog.

If you feel that your choice must come from these breeds, skip ahead to the breed profiles. Mark a big X on the page of each breed listed above. When we start matching your list of desired traits against the breed profiles, you'll compare your list against X-marked breeds instead of against all the breeds. This will save you from going through breeds that don't have the muscle and protective reputation you feel you need.

One final caution: Once you've gone ahead and brought home your protective breed, don't ever allow him to threaten anyone. Let him serve as an imposing presence that will discourage most criminals, but reassure yourself that he will never die because he actually bit a criminal who then called his bluff.

Using the Breed Profiles

Okay, time to find the breeds that match your list of "Traits I'm looking for in a dog."

Start by turning to the breed profiles that begin on page 37. The first profile is that of the Brittany. You'll notice that the profile is divided into three sections. The top section corresponds to the first four categories on your list: Experience Required, With Children, Size, and Coat. The middle section is a chart that corresponds to the final six categories on your list. Right now we'll use just these top and middle sections to find the breeds that best match you.

The first category is Experience Required. Suppose you have written *Fine for novice owners*. Since the Brittany's profile says *Fine for novice owners*, he matches you in this category. Use your pencil to put a check mark on the Brittany's profile, just to the left of *Fine for novice owners*. One category matched.

Next on your list is With Children. If your list says *Good with children*, the Brittany matches you, so you'll put a check mark beside *Good with children* on the Brittany's profile.

Next is Size. Suppose you've listed *Little* and *Small* as your choices. Since the Brittany is *Medium*, he doesn't match you in this category, so you can't put a check mark next to Size.

Do the remaining categories of Coat, Exercise Required, Trimming/Clipping Required, Amount of Shedding, Activity Indoors, Ease of Training, and Sociability with Strangers, putting a check mark on the Brittany's profile beside each category in which he matches you.

When you've compared all ten of your desired traits with all ten Brittany traits, count the check marks, put the total beside the Brittany's name at the top of the page, and move on to the next profile: the Pointer.

You'll notice that some breeds have multiple rankings in a category; for example, a breed may be listed as both high and medium in Ease of Training, meaning some members of the breed are easy to train and some may be a bit stubborn or independent. Even if you've listed only *High* as your choice, you should still consider that the breed matches you in this category. Later on, when we start narrowing your list of suitable breeds, you may decide to eliminate the breed, but for now you can put a check mark beside Ease of Training. Just put a circle around the check mark to indicate that this is not an absolute match.

Yes, combing through all the breeds will take some time, but do it carefully, perhaps over a couple of quiet evenings. You don't want to eliminate a suitable breed because you weren't paying attention to comparing, check-marking, and totaling!

After you've found the total for each profile, turn your list over and write "Suitable Breeds for Me." You're going to list those breeds with the highest scores.

Since the highest score any breed can get is 10, you'll list all these "perfect" breeds first. Depending on how strict your desirable characteristics were, you may have no perfect breeds, or you may have a dozen or more.

Next, go back to any breed that scored 9 and find the nonmatching category. If you feel you could live with a dog who doesn't match you in this category, add the breed to your list. For example, if a 9-point breed doesn't have a check mark next to size because he's small instead of the medium that you wanted, perhaps you'll decide that since he matches you in all the other categories, his small size is forgivable. On the other hand, if he's giant instead of medium, perhaps you'll decide that even though he matches you in all the other categories, you simply don't want a giant dog. Only you know which traits are critical to you and which traits allow room for compromise.

Once you've picked out your 10- and 9-point breeds, you may even want to list 8-point

breeds if your list of "Suitable Breeds for Me" is still very short. When you've compromised as much or as little as you feel comfortable with, take a look at your list. You're going to own one of these breeds!

No? You wouldn't think of owning a Klutz-hound? Why not? Did you have a bad experience with one? Many people approach buying a dog with preconceived notions about various breeds. Perhaps an X-breed growled at you in 1955. Perhaps a Y-breed represented an evil demon in a B movie. Perhaps a friend told you that a Z-breed is the dumbest breed in the world, and he should know because his aunt had one once.

If you're too quick to form an opinion based on a single experience with an X-, Y-, or Z-breed, you're ignoring the hundreds of experiences that were used to create the profiles. If a breed has delighted hundreds of owners, he will delight you, too, as long as he's compatible with you and has been well bred and well raised. Try not to prejudge an entire breed because of a single experience. Try to be open-minded.

But you do have to narrow your list of suitable breeds. We'll use the bottom section of the profiles to narrow your list to three final contestants.

Turn to the profile of your first suitable breed and look at the bottom section, below the chart.

Temperament

The paragraph on temperament describes the breed's personality, behavior, ideal owner or environment, training hints, quirks, and tendencies toward undesirable behaviors like barking or digging. Remember that proper raising and training can modify behavior and prevent or control misbehavior, but the entire purpose of this book is to minimize the need for such extra work by choosing a breed that is compatible to start with.

He is gentle, hardy, and even tempered. He can live in the city if rigorously exercised, but he's better in the suburbs or country with an athletic owner. Too much confinement makes him noisy, hyperactive, and prone to destructive chewing. He is sensitive and very willing to please, so he should be handled gently. He gets along very well with other animals. Some Brittanys are friendly with strangers, some are reserved, some are timid. He should be accustomed to people and noises at an early age. Some whine when excited or nervous.

History

The History paragraph offers a brief history, past and current uses of the breed, and popularity based on AKC statistics.

Once called the Brittany Spaniel, the "spaniel" was dropped because he hunts like a setter—freezing into a point upon scenting a hiding bird. His name comes from the French province of Brittany, where he was the favorite hunting dog of the peasants. Today he is a family hunting dog and companion. AKC popularity: 28th of 130 breeds.

Most of our 175 profiled breeds and varieties are registered with the AKC. They are the largest registry organization in the United States, and they have divided their breeds and varieties into seven groups, based primarily on the original purpose for which each breed was developed:

Sporting Group Breeds that locate and retrieve game birds and water birds (pointers, retrievers, setters, spaniels).

Hound Group Breeds that trail rabbits, foxes, and raccoons by scent (Coonhound) and breeds that locate their prey with their keen sight and run it down with their speed (Afghan, Whippet).

Working Group Breeds that guard livestock (Great Pyrenees), guard homes (Akita), pull carts (Bernese Mountain Dog), pull sleds (Siberian Husky), perform mountain or water rescue (Saint Bernard, Newfoundland), or serve in the military (Doberman).

Terrier Group Breeds that kill rats (Manchester Terrier) and breeds that bolt fox and otter from their dens (Fox Terrier); "terrier" comes from the Latin *terra firma*, meaning "earth."

Toy Group Tiny companion breeds (Pomeranian, Toy Poodle).

Herding Group Breeds that herd sheep or cattle (German Shepherd, Collie, Cardigan and Pembroke Welsh Corgis).

Non-Sporting Group A catchall group of breeds that the AKC felt did not fit in elsewhere (Chow Chow, Dalmatian).

In addition to the seven AKC groups, there are two more groups from which to choose:

Miscellaneous Group Breeds that are growing in popularity but are not yet sufficiently well organized and well distributed in the United States to warrant full recognition (Chinese Sharpei, Chinese Crested); these breeds may be exhibited at AKC dog shows in order to show them and promote them to the public, but they may not win AKC championships.

Rare Breeds Breeds that are not sufficiently well organized or well distributed in the United States to warrant recognition except by their own parent club (Havanese); these breeds may not be exhibited at AKC dog shows.

Physical Features

The Physical Features paragraph describes a good representative of the breed, with specific physical characteristics to seek and avoid.

1. He stands 17½–20½ inches and weighs 30–40 pounds.
2. His coat is slightly wavy, with longer hair on his chest, stomach, legs, and tail, and requires brushing and combing twice a week. Straggly hairs need to be scissored every three months. He is mostly white, with orange or liver (brown) patches and flecks.
3. His ears hang down; when pulled forward, they reach about half the length of his muzzle. His eyes are brown or amber (brownish yellow). His nose is light brown, fawn, or deep pink. His teeth meet in a scissors bite. His tail is a natural bobtail or is docked under 4 inches long.
4. Faults: His nose should not be spotted. His length should not be greater than his height.

5. Disqualifications: His nose should not be black. He should not have black hairs in his coat. His tail should not be over 4 inches long. He should not stand over or under the height specification.

In the information on physical features, the first and second paragraphs, dealing with height and weight, and coat, grooming, and colors are self-explanatory, but the third, fourth, and fifth items need further discussion.

Item 3 describes the breed's ears, eyes, nose, bite, and tail.

Type of ears Ears that hang down fall naturally on both sides of the dog's head (Beagle, Labrador Retriever). Ears that prick up start out hanging in a puppy, but stand up by themselves when the puppy is two to five months old (German Shepherd). Ears that fold forward look like prick ears, but the top part of the ear tips forward (Collie). Ears that fold backward flatten back along the skull (Greyhound). Cropped ears start out hanging, but are surgically cut and taped to stand erect (Doberman).

Surgically cut? Let's discuss the cosmetic surgery of ear cropping, in case you're considering a breed in which it's commonly done. Your veterinarian anesthetizes your puppy, cuts his round hanging ears into a triangular shape, and bandages the ears in an erect position for several weeks. There is some pain or discomfort, depending on the skill of your veterinarian and the sensitivity of your puppy. Often the ears refuse to stand correctly and the puppy is faced with further bandaging or a lifetime of half-standing, half-flopping ears. Today many veterinarians refuse to crop ears, and soon the procedure may be illegal in the United States, as it already is in Europe.

Ear cropping has an interesting history: The short ears gave guarding Dobermans a satanic look, allowed trespassers no swinging ears to grab onto during a fight, increased hearing, because a prick ear is supposedly better able to capture faint sounds, and decreased the likelihood of ear infection, because a prick ear allows air to circulate and cleanse.

But why do we continue to crop ears today? Many breeders say, "Because Dobermans look funny with long ears." Why? Why do we think that Labradors are beautiful with hanging ears, but that Dobermans are not? Fortunately, attitudes are becoming more thoughtful and questioning. Dog owners across America are discovering, to their surprise and delight, just as their European counterparts did, that their Dobermans look just as handsome and will love them and protect them just as earnestly when their natural ears are left alone.

Ear cropping: The choice is yours. (If you plan to show the dog, however, keep in mind that some breeds must be cropped to be judged fairly.)

To return to item 3 under Physical Features:

Color of eyes

Brown eyes are most common, but blue eyes (Siberian Husky), yellow eyes (Chesapeake Bay Retriever), or merle eyes (a swirled mixture of blue/gray/white—Australian Shepherd) are allowed in some breeds. If a standard allows blue eyes, one or both may be blue. If a standard does not allow blue eyes, neither eye may be blue.

Color of nose

A black nose is most common, but a brown nose (Vizsla) or a flesh-colored or pink-streaked nose (Siberian Husky) is allowed in some breeds.

Bite

is the way in which a dog's upper and lower teeth meet. Most common is a scissors bite, where the back of the upper front teeth slightly overlap but still touch the front of the lower front teeth. Also common is a level (or even) bite, where the bottom of the upper front teeth meet the top of the lower front teeth edge to edge. Undesirable in most breeds is an undershot bite, where the lower front teeth jut out beyond the upper front teeth. Undesirable in all breeds is an overshot bite, where the upper front teeth jut out beyond the lower front teeth.

A proper bite is important. Most breeds developed for working purposes used their teeth for nipping livestock, biting trespassers, bringing down game, retrieving fallen birds, and so on. A proper bite allows a dog to chew his food easily, and many dogs with improper bites develop tooth problems. To keep a breed true to its original form, the proper bite must be maintained.

Tail

Most common is a long tail that hangs down but sweeps back and forth when the dog is excited (German Shepherd, Golden Retriever); if the type of tail is not mentioned in a profile, it is simply a long tail. Some breeds wear a curled tail (Alaskan Malamute) or a docked tail (Airedale, Cocker Spaniel, Australian Shepherd). Since a docked tail has been cut short surgically, let's discuss its relationship to ear cropping.

Tail docking: Which breeds and why? Dobermans and Boxers, among others and, again, to give trespassers one less appendage to grab onto. Why then are some bird dogs, like Cocker Spaniels, docked? To protect their tails from thwacking against thickets and fences. Why then are other bird dogs, like English Setters, not docked? Because they tend to work more in wide-open fields than in close-quarter thickets.

If your Boxer is not a guard dog and your Cocker is not a hunting dog, should their tails still be docked? As a buyer you usually don't have much choice. Docking is done by the breeder's veterinarian when the litter is only a few days old. Newborn puppies have little pain sensation, their tails are thin, and when docking is done by a skilled veterinarian, the puppy seems hardly to feel it and the cut heals quickly. Dogs with docked tails obviously enjoy the sensation of tail-wagging just as much as dogs with long tails, and certainly less breakables go flying off the coffee table. So, although tail docking is cosmetic surgery, my own opinion is that it is not inhumane.

Items 4 and 5 describe faults and/or disqualifications.

Faults are traits that are atypical and undesirable for a particular breed. Disqualifications are traits that are *so* atypical and undesirable for a particular breed that a dog with such traits may not even compete in the show ring against others of his breed; he is too different from what his developers intended him to look and act like.

Faults and disqualifications help you to notice whether an adult dog is a good representative of his breed. Soon we'll be searching for your puppy, and when you visit a kennel, you'll use this section of faults and disqualifications to help you judge whether the adult dogs at the kennel are good representatives or poor representatives of your breed. You'll only choose a puppy whose parents are good representatives so that he has the best chance of becoming a good representative himself.

When we listed good traits and poor traits, we chose to concentrate on size, coat, color, head, and tail. The AKC breed standards go into great detail about length of back, angle of shoulder placement, and so on, and these traits are certainly important in distinguishing one breed from another, but they are difficult for a novice—and sometimes even for an expert!—to observe and compare. Size, coat, color, head, and tail are the areas where a novice can observe whether a dog meets his standard and thus is a reasonably typical representative of his breed.

Note: Although they may not be mentioned specifically in each profile, the following traits are faults or disqualifications in *all* breeds: an undershot bite (except in a few breeds), an overshot bite, shyness or aggressiveness towards people, the absence of one or both testicles in males, and any characteristic opposed to those listed in items 1, 2, and 3. In other words, if 1 says that a breed should stand sixteen to twenty inches and you see a member of that breed who stands twenty-four inches, he is faulty, even if "oversized" is not mentioned specifically in the Faults paragraph. We chose to mention specifically only the faults that are most commonly seen.

Health Problems

The Health Problems paragraph lists the medical problems to which the breed is susceptible.

He is susceptible to glaucoma and spinal paralysis.

The most common and most serious health problems are:

Hip dysplasia This disorder, where the hip does not fit properly into its socket, is mostly of genetic (inherited) origin. Theoretically, all breeds are susceptible to this crippling disorder, but it is most commonly seen in medium, large, and giant breeds. Severe cases may show up in young puppies, moderate cases usually appear between six to twelve months of age, and mild cases may not show up until one to two years or even later. Treatment includes major surgery, physical therapy, and/or painkillers.

The best preventative is buying *only* puppies whose parents' hips were X-rayed by a veterinarian and found to be free of the condition before being bred. This does not guarantee that the puppy won't develop the condition, but it decreases the odds immensely. Usually the breeder sends the X-rays to the Orthopedic Foundation of America (OFA) for evaluation and an official clearance certificate. If you are looking at a breed that is listed in the profile as susceptible to hip dysplasia, it is imperative that you ask to see the OFA certificates—or at the very least, the X-rays and a letter of clearance from a veterinarian—for both parents. If the breed is not specifically listed as susceptible, but is medium-sized or bigger, I still recommend that you look for breeders who are taking the precaution of X-raying their dogs' hips. Routine X-raying is the best way to ensure that a breed remains free of this all-too-common hip disorder.

Progressive retinal atrophy (PRA) and hereditary cataracts

These eye diseases are of genetic origin and result in severe loss of vision and usually blindness. Some cases can be detected in six-week-old puppies, but other cases may not show up until five to seven years of age. The best preventative is buying only puppies whose parents' eyes were examined by a canine ophthalmologist and found to be free of hereditary eye conditions before being bred. Usually the breeder sends the tests to the Canine Eye Registry Foundation (CERF) for evaluation and

an official clearance certificate. If you are looking at a breed that is listed in the profile as susceptible to PRA or hereditary cataracts, it is imperative that you ask to see the CERF certificates—or at the very least, a letter of clearance from a board-certified canine ophthalmologist—for both parents. Many breeders, especially of Collies and Shelties, will also have the eyes of each individual puppy examined before selling him.

von Willebrand's Disease

This blood-clotting disorder is of genetic origin and produces external or internal hemorrhaging from a simple cut or illness. The best preventative is buying *only* puppies whose parents' blood was tested and found to be free of the condition before being bred.

Bloat

This is an emergency syndrome of unknown origin that primarily strikes large, deep-chested breeds. The stomach swells with food, water, and/or internal gases, seals itself off from relief, and may suddenly twist or flip over, resulting in death, unless there is immediate surgical intervention. Bloat-susceptible breeds should always be fed several small meals rather than one large meal, and water and exercise should be withheld for an hour or two after meals.

Eyelid or eyelash abnormalities

These are uncomfortable conditions, but they can be corrected by surgery. *Entropion* is the eyelid turning inward and irritating the eyeball. *Ectropion* is the eyelid turning outward. *Distichiasis* is extra eyelashes irritating the eyeball.

Other health problems include skin conditions (allergies, eczema, cysts), eye diseases (cataracts, glaucoma), epilepsy, heart disease, cancer, ear infections, slipped discs, hernias, kidney disease, digestive upsets, congenital deafness, jawbone disorders, slipped joints, diabetes, and respiratory problems. Unfortunately, dogs suffer many of the same ailments we do.

Cautions when Buying

The Cautions when Buying paragraph offers specific advice and warnings when buying this breed. While it is not mentioned specifically in each profile, the caution to "buy from a good breeder" should always be foremost in your thinking. In Step Three, you'll learn how to distinguish clearly between good and poor breeders, and how to locate reputable breeders.

Don't choose a timid or excitable Britt puppy.

Keep in mind that the breed profiles are for *mature* dogs. A dog is physically mature when he has attained his adult height, weight, coat, and color. He is sexually mature when he is capable of breeding. He is emotionally mature when he loses most of his silliness and consistently acts as indicated in the temperament paragraph. Just as with people, physical and sexual maturity usually occurs before emotional maturity. And just as with people, dogs are not puppies one day and adults the next; they go through funny, awkward, gangly adolescent stages, both physically and emotionally.

Small breeds are usually mature at six to twelve months, medium breeds at twelve to eighteen months, and large breeds at eighteen to thirty-six months. So an eleven-month-old Rottweiler is not going to have the solid build, self-confidence, or protectiveness of a mature

Rottweiler. Be especially patient with breeds whose profiles indicate they mature more slowly than other breeds.

Now that you understand how to read the five paragraphs of the bottom section, how should you use them to narrow your list of suitable breeds?

Let's suppose the Brittany is one of your suitable breeds. Look first at his temperament: ". . . He can live in the city if rigorously exercised, but he's better in the suburbs or country with an athletic owner. Too much confinement makes him noisy, hyperactive, and prone to destructive chewing. . . ." If you are a jogger or hunter with an outside yard, this Brittany trait presents no problems. But if you are a sedate person living in a Manhattan studio apartment, you should eliminate the Brittany from further consideration. Draw a line through his name.

Similarly, "He is sensitive and very willing to please, so he should be handled gently . . . some are timid. He should be accustomed to people and noises at an early age." If you are a gentle, sensitive person yourself, fine. But if you have a loud voice, a heavy hand, or if you envision yourself the master of a macho dog, cross off the gentle Brittany.

Similarly, a profile that warns of a breed's possible aggressiveness with other dogs might be cause for concern if there are many dogs in your area. If you have cats or rabbits, you would of course be cautious with a breed that is listed as suspect with these small pets, but also be aware that all dogs should be carefully introduced to small pets and that breeds who are listed as aggressive with other *dogs* should be more carefully introduced to small pets than breeds who are listed as fine with all other animals.

Read through the rest of the profile. If you see anything that makes you shake your head, wrinkle your nose, or otherwise feel doubtful about having this dog in your home as part of your family, eliminate him from further consideration. Remember, based on ten important traits, these breeds are all potentially good matches for you, so you can base further narrowing on anything you want: body shape (the pudgy Bulldog, the low-slung Dachshund, the streamlined Whippet), color, type of ears, type of tail, type of health problems.

Your goal is to keep just three profiles—three final contestants.

5

Choosing Your Breed

Time to make your final decision! If you're stuck between a couple of breeds, you may need to visit some kennels and/or dog shows to get an up-close look at the dogs. We'll be doing that soon, so go through the next step doing everything in duplicate, and somewhere along the line one of the breeds will stand out as the best one for you.

Also, check bookstores and libraries for a picture book on the breeds you're stuck between—books such as, *This is the Keeshond, The Complete Borzoi, How to Raise and Train a Puli,* or *Your Papillon.* You'll probably find books for the most popular AKC breeds; unfortunately, you probably won't find many for the less-common AKC breeds or for the non-AKC breeds.

Pet shops also sell books, but *don't* yield to the temptation to buy a puppy while you're there. In fact, don't even think that a pet shop is a good place to get a look at the breed you're in-

terested in. You'll soon learn that a pet shop's Collie is *not* the same dog at all as a breeder's Collie. Be patient. Your puppy is coming!

As you're looking at pictures of the breeds, don't hesitate to make your decision based on appearance. If several breeds have temperaments and behaviors that fit what you're looking for, choose the breed that is most attractive to you. Problems arise when people select a breed based *only* on appearance, though. Remember that beauty is only skin deep. If you choose a beautiful breed whose personality and behavior don't suit you, you're going to spend a lot of time disliking the dog and you'll find his beauty fading in your eyes. If, on the other hand, you choose a breed with a personality that appeals to you but an appearance that doesn't, you'll be astonished, as time goes by, at how beautiful he becomes in your eyes.

So make your final choice! Then we'll start looking for your breeder.

The Breed Profiles

The Sporting Group

Brittany · Pointer · German Shorthaired Pointer

German Wirehaired Pointer · Chesapeake Bay Retriever

Curly-Coated Retriever · Flat-Coated Retriever · Golden Retriever

Labrador Retriever · English Setter · Gordon Setter · Irish Setter

American Water Spaniel · Clumber Spaniel · American Cocker Spaniel

English Cocker Spaniel · English Springer Spaniel · Field Spaniel

Irish Water Spaniel · Sussex Spaniel · Welsh Springer Spaniel · Vizsla

Weimaraner · Wirehaired Pointing Griffon

Brittany 5

Fine for Novice Owners ✓
Good with Children ✓
Medium in Size
Feathered, Medium-Length Coat

	High	Med	Low
EXERCISE REQUIRED	●		
TRIMMING/CLIPPING REQUIRED ✓		●	
AMOUNT OF SHEDDING		●	
ACTIVITY INDOORS ✓		●	
EASE OF TRAINING ✓	●		
SOCIABILITY WITH STRANGERS		●	

Temperament

He is gentle, hardy, and even tempered. He can live in the city if rigorously exercised, but he's better in the suburbs or country with an athletic owner. Too much confinement makes him noisy, hyperactive, and prone to destructive chewing. He is sensitive and very willing to please, so he should be handled gently. He gets along very well with other animals. Some Brittanys are friendly with strangers, some are reserved, some are timid. He should be accustomed to people and noises at an early age. Some whine when excited or nervous.

History

Once called the Brittany Spaniel, the "spaniel" was dropped because he hunts like a setter—freezing into a point upon scenting a hiding bird. His name comes from the French province of Brittany, where he was the favorite hunting dog of the peasants. Today he is a family hunting dog and companion. AKC popularity: 28th of 130 breeds.

Physical Features

1. He stands 17½–20½ inches and weighs 30–40 pounds.
2. His coat is slightly wavy, with longer hair on his chest, stomach, legs, and tail, and requires brushing and combing twice a week. Straggly hairs need to be scissored every three months. He is mostly white, with orange or liver (brown) patches and flecks.
3. His ears hang down; when pulled forward, they reach about half the length of his muzzle. His eyes are brown or amber (brownish yellow). His nose is light brown, fawn, or deep pink. His teeth meet in a scissors bite. His tail is a natural bobtail or is docked under 4 inches long.
4. Faults: His nose should not be spotted. His length should not be greater than his height.
5. Disqualifications: His nose should not be black. He should not have black hairs in his coat. His tail should not be over 4 inches long. He should not stand over or under the height specification.

Health Problems

He is susceptible to glaucoma and spinal paralysis.

Cautions when Buying

Don't choose a timid or excitable Britt puppy.

Pointer

For Experienced Owners Only
Good with Children if Raised with Children
Large in Size
Smooth Coat

	High	Med	Low
EXERCISE REQUIRED	🦴		
TRIMMING/CLIPPING REQUIRED			🦴
AMOUNT OF SHEDDING		🦴	
ACTIVITY INDOORS		🦴	
EASE OF TRAINING		🦴	
SOCIABILITY WITH STRANGERS		🦴	

Temperament

If raised as a house pet from puppyhood, he is dignified, extremely sweet, and gentle. But he is also packed with nervous energy and in need of running exercise—he's unsuited to the city. Too much confinement makes him noisy, restless, and prone to destructive chewing. He has a tendency toward hyperactivity and nervousness—he should live with a quiet, calm (but athletic) family who will keep him relaxed and exercised. Most are slightly reserved or timid with strangers, so he should be accustomed to people and noises at an early age. He gets along well with other animals. A bit stubborn and easily distracted, he does respond to patient obedience training that does not include a lot of jerking around. Commands such as *down* and *stay* are important in controlling this breed. Some Pointers raised as kennel dogs can be high-strung.

History

He freezes into a point when he scents a hiding bird, and when his owner has shot it, he retrieves. Developed in England, often called the English Pointer, he is probably the oldest of the pointing breeds. Spectacularly fast and wide-ranging, with a concentration and competitiveness far surpassing other sporting dogs, he is the king of national pointing trials—the English Setter is the only breed that can keep up with him. He is a family hunting dog and companion. AKC popularity: 95th of 130 breeds.

Physical Features

1. Males stand 25–28 inches, females stand 23–26 inches. He weighs 45–75 pounds.
2. He has a hard coat, which needs only a quick brush once a week. He is mostly white, with black, orange, lemon, or liver (brown) patches, speckles, and flecks. Solid black and solid liver are less common.
3. His ears hang down to just below his lower jaw. His eyes are brown with an intense expression. His nose is black on black-patched dogs and brown on liver-patched dogs, but may be lighter on orange- and lemon-patched dogs. His teeth meet in a scissors or level bite.

Health Problems

He is susceptible to hip dysplasia, dwarfism, thyroid problems, and skin conditions. Buy only from OFA-registered parents.

Cautions when Buying

Pointers are divided into field types and show types. A Pointer from show lines is bigger and heavier, and you shouldn't expect him to work like a Pointer from field lines. The Pointers that make the best pets tend to come from the show lines; field Pointers are often too businesslike and energetic to be at their best as pets. Don't choose the most independent or energetic Pointer puppy or a timid puppy.

German Shorthaired Pointer

For Experienced Owners Only
Good with Older, Considerate Children
Large in Size
Smooth Coat

	High	Med	Low
EXERCISE REQUIRED	🦴		
TRIMMING/CLIPPING REQUIRED			🦴
AMOUNT OF SHEDDING		🦴	
ACTIVITY INDOORS	🦴		
EASE OF TRAINING			🦴
SOCIABILITY WITH STRANGERS		🦴	

Temperament

He is good-natured and dignified, but so exuberant and in need of running exercise that he is unsuited to the city. Too much confinement makes him noisy, restless, and prone to destructive chewing. Some have a tendency toward hyperactivity and nervousness—he should live with a quiet, calm (but athletic) family who will keep him relaxed and exercised. Some are friendly with strangers, some are reserved, some are timid, some are protective—so he should be accustomed to people and noises at an early age. He likes children, but may be too energetic for toddlers. He can be aggressive with strange dogs. Strong-willed and easily distracted, he needs firm, persistent obedience training. Commands such as *down* and *stay* are important in controlling this breed. Some can be excitable, some bark a lot, and some can be hard to housebreak.

History

He freezes into a point when he scents a hiding bird, and when his owner has shot it, he retrieves. He was developed in Germany as a rugged hunter of birds, ducks, rabbits, and raccoons, and he has both bird dogs and scenthounds in his ancestry. He is a family hunting dog and companion. AKC popularity: 35th of 130 breeds.

Physical Features

1. Males stand 23–25 inches, females stand 21–23 inches. He weighs 45–70 pounds.
2. He has a hard coat, which needs only a quick brushing once a week. He is liver (brown) with much gray and white flecking. Solid liver is less common.
3. His ears hang down. His eyes are dark with a good-humored expression. His nose is brown. His teeth meet in a scissors bite. His tail is docked.
4. Faults: His tail should not curve toward his head.
5. Disqualifications: His eyes should not be wall eyes. His nose should not be flesh colored. His teeth should not meet in an extremely overshot or undershot bite. He should not have black, red, orange, lemon, or tan hairs in his coat. He should not be solid white.

Health Problems

He is susceptible to hip dysplasia. Buy only from OFA-registered parents.

Cautions when Buying

Don't choose the most boisterous or independent Shorthair puppy, and don't choose a timid or excitable puppy.

German Wirehaired Pointer

For Experienced Owners Only
Good with Older, Considerate Children
Large in Size
Wiry Coat

	High	Med	Low
EXERCISE REQUIRED	✦		
TRIMMING/CLIPPING REQUIRED		✦	
AMOUNT OF SHEDDING		✦	
ACTIVITY INDOORS	✦		
EASE OF TRAINING			✦
SOCIABILITY WITH STRANGERS		✦	

Temperament

He has a steady, sensible, dignified personality, but he is also rugged, determined, and in need of running exercise—he's unsuited to the city. Too much confinement makes him noisy, restless, and prone to destructive chewing. Reserved and protective with strangers, he should be accustomed to people at an early age. He can be aggressive with other dogs. He is stubborn and bold, and requires firm, persistent obedience training. This breed is usually more serious and more discriminating than his Shorthair cousin.

History

He freezes into a point when he scents a hiding bird, and when his owner has shot it, he retrieves. He was developed in Germany as a rugged-coated hunter in field, forest, thicket, and water. (The Germans have always preferred all-purpose hunters who could handle any type of terrain or species.) Because his ancestry does not include hounds, he is faster than his Shorthair cousin. He is a family hunting dog and companion. AKC popularity: 66th of 130 breeds.

Physical Features

1. Males stand 24–26 inches, females stand 22–24 inches. He weighs 50–70 pounds.
2. His coat is up to 2 inches long and harsh, with bushy eyebrows and a beard, and requires brushing and combing twice a week. Straggly hairs need to be scissored and dead hairs stripped out twice a year. He is liver (brown) with much white flecking. Solid liver is less common. His head is always brown, often with a white blaze.
3. His ears hang down. His eyes and nose are brown. His teeth meet in a scissors bite. His feet are webbed to aid in swimming. His tail is docked.
4. Faults: His nose should not be flesh colored or spotted. He should not have black hairs in his coat. His coat should not be much over 2 inches long.

Health Problems

He is susceptible to hip dysplasia. Buy only from OFA-registered parents.

Cautions when Buying

Don't choose the boldest, most active, or most independent Wirehair puppy, and don't choose a timid puppy.

Chesapeake Bay Retriever

For Experienced Owners Only
Good with Children
Large in Size
Short Coat

	High	Med	Low
EXERCISE REQUIRED		🦴	🦴
TRIMMING/CLIPPING REQUIRED			🦴
AMOUNT OF SHEDDING		🦴	🦴
ACTIVITY INDOORS			🦴
EASE OF TRAINING		🦴	🦴
SOCIABILITY WITH STRANGERS			🦴

Temperament

He is the most rugged and powerful of the five retriever breeds. This calm, sensible, capable dog can do well in the city if given long daily walks, occasional runs, and whenever possible, swimming and fetching. Reserved and protective with strangers, he should be accustomed to people at an early age. He can be aggressive with other dogs. Although stubborn, he does respond well to obedience training that is firm and consistent. If not well trained, he can refuse commands from family members who have not established leadership over him.

History

One of the few breeds developed in the United States, around the icy bays of Maryland, he is one of the greatest retrievers of ducks and geese, especially in rough, frigid waters. He was once the king of national retriever trials, but the faster Labrador now holds that crown. He is a family hunting dog and companion. AKC popularity: 45th of 130 breeds.

Physical Features

1. Males stand 23–26 inches, females stand 21–24 inches. He weighs 55–80 pounds.
2. His coat is coarse and slightly wavy, especially along his back, and needs to be brushed once a week. He is deep brown, faded tan, or straw colored (called dead-grass). His adult color becomes evident around eleven weeks of age. A little white on his chest, toes, and belly is okay.
3. His ears hang down. His eyes are yellowish or amber (brownish yellow). His nose is light brown to flesh colored. His teeth meet in a scissors or level bite. His feet are webbed to aid in swimming.
4. Disqualifications: His coat should not be curly, nor over 1¼ inches long. He should not have white hairs in his coat (except where allowed above). He should not be solid black. His teeth should not meet in an overshot or undershot bite. He should not have dewclaws on his hind legs.

Health Problems

He is susceptible to hip dysplasia, PRA, cataracts, and eczema. Buy only from OFA- and CERF-registered parents.

Cautions when Buying

Some breeders strive for working ability and Standard conformation in the same dog, but some breeders strive for one or the other. The two types may or may not look different, but you shouldn't expect a Chesapeake from show lines to work like a Chesapeake from field lines. The Chesapeakes that make the best pets tend to come from the dual (field and show) or show lines; field Chesapeakes are sometimes too determined and businesslike to be at their best as pets. Don't choose the boldest or most independent Chessie puppy.

Curly-Coated Retriever

Fine for Novice Owners
Good with Children
Large in Size
Curly Coat

	High	Med	Low
EXERCISE REQUIRED		●	●
TRIMMING/CLIPPING REQUIRED			●
AMOUNT OF SHEDDING		●	●
ACTIVITY INDOORS			●
EASE OF TRAINING		●	●
SOCIABILITY WITH STRANGERS			●

Temperament

He's quiet indoors, but his daily exercise requirement (including swimming, if possible) is such that he does best in the surburbs or country with someone who will take him out to run. Reserved and sometimes timid or distrustful with strangers, he must be accustomed to people and noises at an early age. He gets along well with other animals. Although very smart, he often uses his intelligence in a clever way that suits his own purposes; thus he needs early, consistent obedience training. However, you must never be harsh with this gentle, sensitive breed, because he can be made skittish and defensive by heavy-handed leash jerking. A Curly is not simply a Golden or Labrador retriever with dark curly hair but rather, an athletic, observant, discriminating dog who thrives on activity, training, and work.

History

The oldest of the five retrievers, he was developed in England to retrieve shot ducks from the water. He is a popular duck and quail retriever in New Zealand and deserves much greater popularity in America, but with so many Labradors and Goldens around, most people have never even heard of him. He is a family hunting dog and companion. AKC popularity: 116th of 130 breeds.

Physical Features

1. He stands 23–27 inches and weighs 60–95 pounds.
2. His coat is a mass of short, crisp curls, including his ears and tail but excluding his head, hocks, and the front of his legs, which are smooth. His coat needs brushing twice a week. He is solid black or solid liver (brown). A few white hairs are allowed on his chest.
3. His ears hang down. His eyes are dark. His nose is black on black dogs, brown on liver dogs. His teeth meet in a scissors or level bite.
4. Faults: He should not have bald patches or uncurled hair on his body.

Health Problems

He is susceptible to hip dysplasia and PRA. Buy only from OFA- and CERF-registered parents.

Cautions when Buying

Don't choose a timid Curly-Coat puppy.

Flat-Coated Retriever

Fine for Novice Owners
Good with Children
Large in Size
Feathered Medium-Length Coat

	High	Med	Low
EXERCISE REQUIRED		▰	
TRIMMING/CLIPPING REQUIRED		▰	
AMOUNT OF SHEDDING		▰	
ACTIVITY INDOORS			▰
EASE OF TRAINING	▰		
SOCIABILITY WITH STRANGERS	▰		

Temperament

He's a cheerful, gentle, reliable tail wagger who needs companionship and attention. He can live in the city if given plenty of daily exercise and, whenever possible, swimming and fetching. This gregarious dog likes people, but he's not as blindly enthusiastic with strangers as is the Golden Retriever. An occasional dog may be a bit timid. He gets along well with other animals. This sweet, willing breed responds very well to obedience training. His only real behavioral problem may include overexuberance and jumping on people.

History

Developed in England to retrieve shot ducks from the water, he was once called the Wavy-Coated Retriever and was immensely popular until the coming of Labradors and Goldens. He deserves greater popularity in America, but with so many Labs and Goldens around, most people have never even heard of him. He is a family hunting dog and companion. AKC popularity: 97th of 130 breeds.

Physical Features

1. He stands 22–24½ inches and weighs 60–70 pounds.
2. His coat is straight or wavy, with longer hair on his chest, stomach, legs, and tail, and needs to be brushed and combed twice a week. Straggly hairs should be scissored every three months. He is solid black (most common) or solid liver (brown).
3. His ears hang down. His eyes are brown or hazel. His nose is black on black dogs, brown on liver dogs. His teeth meet in a scissors or level bite.
4. Faults: His eyes should not be yellow.
5. Disqualifications: His coat should not be yellow or cream colored.

Health Problems

He is susceptible to hip dysplasia and PRA. Buy only from OFA- and CERF-registered parents.

Cautions when Buying

Don't choose the most energetic Flattie puppy, and don't choose a timid puppy.

Golden Retriever

Fine for Novice Owners
Good with Children
Large in Size
Feathered, Medium-Length Coat

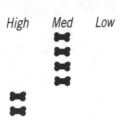

	High	Med	Low
EXERCISE REQUIRED		✦	
TRIMMING/CLIPPING REQUIRED		✦	
AMOUNT OF SHEDDING		✦	
ACTIVITY INDOORS		✦	
EASE OF TRAINING	✦		
SOCIABILITY WITH STRANGERS	✦		

Temperament

This is one of the finest family dogs in the world. He's a cheerful, outgoing, demonstrative dog who does well in the city if given daily exercise and allowed to carry sticks and fetch balls. He likes people and other animals. Some tend to timidity, so he should be accustomed to people and noises at an early age. Although eager to please and wonderfully responsive to obedience training, he's easily distracted by exciting sights and sounds, so you must be calmly persistent. *Never* be harsh with this sweet, gentle dog. His only behavioral problems may include overexuberance and jumping on people. Commands such as *heel, sit, down,* and *stay* are important in teaching him good manners. Also be aware that a Golden is much stronger than he looks—unless he's taught not to pull on the leash, you'll need good biceps to walk him.

History

He was developed in England and Scotland to retrieve shot ducks from the water, but because of his popularity as a family pet, his hunting skills have been all but lost. He is a guide for the blind, a therapy dog for children and the handicapped, and a companion. AKC popularity: 5th of 130 breeds.

Physical Features

1. Males stand 23–24 inches, females stand 21½–22½ inches. He weighs 55–75 pounds.
2. His coat is over 2 inches long, with longer hair on his chest, stomach, legs, and tail, and needs brushing and combing twice a week. Straggly hairs should be scissored every three months. His color ranges from dark red to medium gold to light cream. A few white hairs on his chest are okay. A puppy's coat darkens to its adult color by one year of age.
3. His ears hang down. His eyes are brown, with a kindly expression. His nose is black or brown, but may fade to a lighter color during cold weather. His teeth meet in a scissors bite.
4. Disqualifications: His teeth should not meet in an overshot or undershot bite. He should not have eyelid or eyelash disorders. He should not stand more than 1 inch over or under the height specification.

Health Problems

He is susceptible to hip dysplasia, PRA, von Willebrand's Disease, serious heart problems, cataracts, skin conditions, eyelid/eyelash/retinal abnormalities, and possibly epilepsy. Buy only from OFA- and CERF-registered parents whose hearts have been cardio-examined for soundness. VWD-tested (von Willebrand's Disease) is an additional plus.

Cautions when Buying

THERE ARE MANY GOLDENS AROUND—BE VERY CAREFUL. If you buy this breed from a poor breeder, you could end up with an impossibly flaky, hyperactive Golden. Don't choose a wildly energetic or excitable Golden puppy, and don't choose a timid puppy.

Labrador Retriever

Fine for Novice Owners ✓
Good with Children ✓
Large in Size
Short Coat ✓

	High	Med	Low
EXERCISE REQUIRED ✓		🦴	🦴
TRIMMING/CLIPPING REQUIRED ✓			🦴
AMOUNT OF SHEDDING		🦴	
ACTIVITY INDOORS ✓		🦴	
EASE OF TRAINING ✓	🦴		
SOCIABILITY WITH STRANGERS ✓	🦴		

Temperament

This is another of the finest family dogs in the world. He is so steady, good-natured, and adaptable that he does fine in the city if given vigorous daily exercise and allowed to carry sticks and fetch balls. Some can be destructive if left alone too much. He is usually more conservative and more independent than the Golden Retriever. Most are friendly with strangers, some are slightly reserved. He gets along well with other animals. He's willing to please and very responsive to obedience training, but he's spirited, so he should be trained by a strong owner who will work calmly and consistently with him. Commands such as *heel*, *down*, and *stay* are important in keeping him from becoming too rambunctious. Some Labs have necks like bulls and barely notice tugs on the leash.

History

He originated in Canada, but in the province of Newfoundland—not Labrador. He was fully developed by the British as a water retriever. With his speed and drive, he is now the king of national retriever trials. He is also a family hunting dog, a guide for the blind, a narcotics detection dog, and a companion. AKC popularity: 3rd of 130 breeds.

Physical Features

1. He stands 21½–24½ inches and weighs 55–75 pounds.
2. His coat is straight and dense and needs brushing once a week. He is black, yellow (dark golden or light cream), or chocolate (dark brown). A little white on his chest is allowed. *Note:* Yellow Labs are *not* called golden Labs.
3. His ears hang down. His eyes are dark but may be yellow in chocolate dogs. His nose is black on black and yellow dogs, brown on chocolate dogs, but may fade to pinkish during cold weather. His teeth meet in a level bite. His tail is covered thickly with short hair to give a rounded appearance described as an "otter tail."

Health Problems

He is susceptible to hip dysplasia, PRA, cataracts, eyelid/retinal abnormalities, and bloat. Buy only from OFA- and CERF-registered parents.

Cautions when Buying

THERE ARE MANY LABRADORS AROUND—BE VERY CAREFUL. Labs are divided into field types and show types. A Lab from show lines is heavier and more thickly built, and you shouldn't expect him to work like a Lab from field lines. The Labs that make the best pets tend to come from the dual (field and show) or show lines; field Labs are often too packed with energy and drive to be at their best as pets. If you buy this breed from a poor breeder, you could end up with a tall, rangy, hyperactive Labrador. Don't choose the boldest, most independent, or most energetic Lab puppy, and don't choose a timid puppy.

English Setter

Fine for Novice Owners
Good with Children
Large in Size
Feathered Medium-Length Coat

	High	Med	Low
EXERCISE REQUIRED	🦴		
TRIMMING/CLIPPING REQUIRED		🦴	
AMOUNT OF SHEDDING		🦴	
ACTIVITY INDOORS			🦴
EASE OF TRAINING		🦴	
SOCIABILITY WITH STRANGERS	🦴		

Temperament

He is the mildest-mannered of the three setters. Although quiet and gentle in the house, he's lively outdoors and needs enough running exercise that he belongs in the suburbs or country. He's a sweet dog who gets along well with strangers and other animals, but he's also stubborn and in need of early, firm obedience training. Interestingly, his stubbornness takes the form of resistance rather than wild disobedience: He'll simply brace his legs and refuse to walk. You must be patient and persuasive, never harsh or heavy-handed. He picks up bad habits easily and loses them with difficulty, so don't let him beg from the table even once. He can be hard to housebreak. Exercise and calm, consistent handling are essential with all setters, and setters are happiest around people; they do not thrive as well in homes where they are left alone all day.

History

He freezes into a point when he scents a hiding bird, and when his owner has shot it, he retrieves. He was developed in England and is one of the oldest of the pointing breeds. Fast-working and wide-ranging, he and the English Pointer are the top winners of national pointing trials. He is also a family hunting dog and companion. AKC popularity: 80th of 130 breeds.

Physical Features

1. Males stand 24–25 inches and weigh 60–70 pounds. Females stand 21–23 inches and weigh 45–55 pounds.
2. His coat is straight, with longer hair on his chest, stomach, legs, and tail, and needs brushing and combing twice a week. Straggly hairs should be scissored every three months. Some electric clipping twice a year makes him look extra sharp. He's mostly white, with much flecking and speckling (not usually patches) of black, orange, lemon, or liver.
3. His ears hang down. His eyes are dark with a mild expression. His nose is black or brown. His teeth meet in a scissors or level bite.

Health Problems

He is susceptible to hip dysplasia. Buy only from OFA-registered parents.

Cautions when Buying

English Setters are divided into field types and show types. An English Setter from field lines is smaller, lighter, and has a much shorter coat (often with large patches of color). Either type from a good breeder usually makes a fine pet, but field-bred setters are more active and require more exercise. Don't choose the most independent English Setter puppy.

Gordon Setter

Fine for Novice Owners
Good with Children
Large in Size
Feathered Medium-Length Coat

	High	Med	Low
EXERCISE REQUIRED	●		
TRIMMING/CLIPPING REQUIRED		●	
AMOUNT OF SHEDDING		●	
ACTIVITY INDOORS		●	
EASE OF TRAINING		●	
SOCIABILITY WITH STRANGERS		●	

Temperament

He is the most serious and sensible of the three setters. He's an athletic, energetic dog who belongs in the suburbs or country with someone who will take him out to run. Too much confinement makes him hyperactive and rambunctious. Although usually sweet tempered, he can be demanding of attention and childishly jealous of other pets. He's reserved and sometimes protective with strangers and can be aggressive with strange dogs. His stubbornness calls for early, firm obedience training, but be patient and persuasive, never harsh—all setters are sensitive. He picks up bad habits easily and loses them with difficulty, so don't let him beg from the table even once. He can be hard to housebreak. Exercise and calm, consistent handling are essential with setters.

History

He freezes into a point when he scents a hiding bird, and when his owner has shot it, he retrieves. He was developed in Scotland, his name coming from the Scottish Duke of Gordon. Daniel Webster imported the first Gordons to America. He is a family hunting dog and companion. AKC popularity: 63rd of 130 breeds.

Physical Features

1. He stands 23–27 inches and weighs 45–80 pounds.
2. His coat may be straight or wavy, with longer hair on his chest, stomach, legs, and tail, and needs brushing and combing twice a week. Straggly hairs should be scissored every three months, and electric clipping twice a year makes him look extra sharp. He is black, with rich tan markings above his eyes, on his muzzle, throat, chest, legs, feet, and under his tail. Narrow black stripes on his toes (called penciling) and a little white on his chest are allowed.
3. His ears hang down. His eyes are dark brown with a wise expression. His nose is black. His teeth meet in a scissors or level bite.
4. Disqualifications: He should not be mostly tan or red.

Health Problems

He is susceptible to hip dysplasia. Buy only from OFA-registered parents.

Cautions when Buying

Unlike most sporting breeds, many Gordons are bred to be dual-purpose (both hunting and show). A Gordon from strict field lines is usually smaller, lighter, and has a shorter coat than a Gordon from strict show lines, but makes just as good a pet. Don't choose the most independent Gordon puppy.

Irish Setter

Fine for Novice Owners
Good with Children
Large in Size
Feathered Medium-Length Coat

	High	Med	Low
EXERCISE REQUIRED	●		
TRIMMING/CLIPPING REQUIRED		●	
AMOUNT OF SHEDDING		●	
ACTIVITY INDOORS	●		
EASE OF TRAINING	●	●	
SOCIABILITY WITH STRANGERS	●		

Temperament

He has been described as happy-go-lucky, rollicking, bold, clownish, impulsive, flighty, high-strung, demonstrative, and/or excitable. Some are more dignified and reserved, and these are impressive and aristocratic examples of the breed. He needs running exercise and is much less of an energetic handful in the suburbs or country. Too much confinement and too little exercise make him hyperactive and destructive. He gets along well with strangers and other animals. He's a willing dog who responds well to early, firm (but as with all sensitive setters, never harsh) obedience training, but without sufficient time and effort, he is far too rambunctious. He picks up bad habits easily and loses them with difficulty, so don't let him beg from the table even once. He can be hard to housebreak. Exercise, ongoing obedience work, and calm, consistent handling are essential with an Irish.

History

He freezes into a point when he scents a hiding bird, and when his owner has shot it, he retrieves. He was developed in Ireland and originally was red and white; Irish Setters in Europe can still be that color. Today he is primarily a companion, very occasionally a family hunting dog. AKC popularity: 52nd of 130 breeds.

Physical Features

1. Males stand 26–28 inches, females stand 24–26 inches. He weighs 60–70 pounds.

2. His coat is silky, with longer hair on his chest, stomach, legs, and tail, and needs brushing and combing twice a week. Straggly hairs should be scissored every three months, and electric clipping twice a year makes him look extra sharp. He ranges from mahogany red to chestnut red (puppies are lighter), with some white allowed on his chest, throat, toes, and as a narrow streak on his forehead.

3. His ears hang down; when pulled forward, they nearly reach his nose. His eyes are brown. His nose is black or brown. His teeth meet in a scissors or level bite.

Health Problems

He is susceptible to hip dysplasia, PRA, bloat, congenital heart disease, skin conditions, and epilepsy. Buy only from OFA- and CERF-registered parents.

Cautions when Buying

Irish are divided into field types and show types. An Irish from field lines is smaller, lighter, and has a much shorter coat—some are red and white. You shouldn't expect an Irish from show lines to work like an Irish from field lines (though there are few hunting Irish today), but either type from a good breeder usually makes a fine pet. Don't choose the most active, boisterous, independent, or excitable Irish puppy, and don't choose a timid puppy.

American Water Spaniel

Fine for Novice Owners
Good with Older, Considerate Children
Small to Medium Size
Curly Coat

	High	Med	Low
EXERCISE REQUIRED		●	
TRIMMING/CLIPPING REQUIRED			●
AMOUNT OF SHEDDING		●	
ACTIVITY INDOORS		●	
EASE OF TRAINING	●		
SOCIABILITY WITH STRANGERS		●	

Temperament

He is hardy, enthusiastic, and easy to get along with. Because he's adaptable and asks only to be included in activities, he can do well in the city if given daily exercise and, whenever possible, swimming and fetching. Most are mildly friendly with strangers. Some are timid, so he should be accustomed to people and noises at an early age. Some can be scrappy with strange dogs. This pleasant, willing dog is highly sensitive, so obedience training should be calm, quiet, and persuasive, rather than demanding or sharp. He could snap defensively if startled or harshly disciplined. Some bark and whine a lot, and some drool.

History

He can work like a spaniel by flushing birds or small game from their hiding places, or he can retrieve shot ducks from the water, his slender tail acting as a rudder as he swims. He is one of the few breeds to have been developed in America, around the icy bays of Wisconsin, by hunters who wanted a durable, hard-working, mobile hunter who would not tip the boat over when he climbed in and who could withstand Wisconsin's frigid waters. This versatile little dog deserves much greater popularity, but with the Cocker and Springer spaniels around, most people have never even heard of him. He is a family hunting dog and companion. He is also the state dog of Wisconsin. AKC popularity: 96th of 130 breeds.

Physical Features

1. He stands 15–18 inches and weighs 24–45 pounds.
2. His coat is a mass of short, crisp curls, including his ears and tail but excluding his head, which is very smooth. His coat needs brushing twice a week. He is solid liver (medium brown) or solid chocolate (dark brown). A little white on his chest and toes is allowed.
3. His ears hang down; when pulled forward, they reach his nose. His eyes are brown or hazel. His nose is brown. His teeth meet in a scissors or level bite.
4. Faults: His tail should not be smooth and ratlike; that trait belongs to the Irish Water Spaniel, not the American.
5. Disqualification: His eyes should not be yellow.

Health Problems

He is susceptible to some skin conditions.

Cautions when Buying

Don't choose the most independent Water Spaniel puppy, and don't choose a timid puppy.

Clumber Spaniel

Fine for Novice Owners
Good with Children
Medium in Size
Feathered Medium-Length Coat

	High	Med	Low
EXERCISE REQUIRED		●	
TRIMMING/CLIPPING REQUIRED		●	
AMOUNT OF SHEDDING		●	
ACTIVITY INDOORS			●
EASE OF TRAINING		●	
SOCIABILITY WITH STRANGERS		●	

Temperament

He is dignified, steady, almost imperturbable. He's quiet indoors and fine in the city, but he does need exercise; although he might prefer just to lie around, he's a heavy breed and must be taken out regularly to stay fit. He's often a one-person dog, reserved with strangers and faithful to his owner in a stolid, undemonstrative way. He gets along well with other animals. Although stubborn, he does respond to patient, firm, persuasive repetition of obedience commands, but because of his easygoing approach to life, he's seldom a problem even when he doesn't obey very quickly. Overall, he's not a playful canine pal, but a calm companion. He drools, snores, and wheezes.

History

He flushes birds from their hiding places and then retrieves them when the hunter shoots. His name comes from Clumber Park in England, but he was also developed partly in France, and a now-extinct Alpine spaniel may be in his ancestry. Although he was one of the first ten breeds recognized by the AKC, he is now one of the ten rarest. He is primarily a companion, very occasionally a family hunting dog. AKC popularity: 123rd of 130 breeds.

Physical Features

1. Males stand 19–20 inches and weigh 70–85 pounds. Females stand 17–19 inches and weigh 55–70 pounds.
2. His coat is soft, with longer hair on his chest, stomach, legs, and tail, and needs brushing and combing twice a week. Straggly hairs should be scissored every three months. He's mostly white, with lemon or orange markings on his ears, around his eyes, near his tail, and/or as freckles on his muzzle and front legs.
3. His ears hang down. His eyes are amber (brownish yellow) with a dignified expression. His nose is various shades of brown, beige, and/or rose. His teeth meet in a scissors bite. His tail is docked.

Health Problems

He is susceptible to hip dysplasia and entropion.

Cautions when Buying

Don't choose the most independent Clumber puppy.

American Cocker Spaniel

Fine for Novice Owners
Good with Children
Small in Size
Feathered Medium-Length Coat

	High	Med	Low
EXERCISE REQUIRED		▰	
TRIMMING/CLIPPING REQUIRED	▰		
AMOUNT OF SHEDDING		▰	
ACTIVITY INDOORS		▰	
EASE OF TRAINING	▰		
SOCIABILITY WITH STRANGERS	▰		

Temperament

The Cocker of good breeding is lively, cheerful, gentle, and playful. He does fine in the city if exercised regularly and if accustomed to people and noises at an early age, for some can be a bit timid. He gets along well with strangers and other animals. He's a willing dog, very responsive to persuasive obedience training—never hit this sensitive breed or he can become snappy and defensive. Some bark a lot, and some can be hard to housebreak. The Cocker of poor breeding can be nasty, with bizarre physical and behavioral problems.

History

He flushes birds from their hiding places and then retrieves them when the hunter shoots. He was developed in England (not in America) as a lively little hunter, and his name comes from the woodcock game bird. He is now so popular as a show dog and companion that his hunting ability has been all but lost. AKC popularity: 1st of 130 breeds.

Physical Features

1. Males stand 14½–15½ inches, females stand 13½–14½ inches. He weighs 24–28 pounds.
2. His coat is silky, with longer hair on his chest, stomach, legs, and tail, and needs brushing and combing twice a week. He needs scissoring and electric clipping every three months. Colors include solid buff (most common), solid black, parti-color (black-and-tan, white-and-tan), tricolor (black/tan/white), and others. Some white on his chest or throat is okay.
3. His ears hang down; when pulled forward, they reach at least to his nose. His eyes are brown or hazel. His nose is black on black dogs, possibly brown on others. His teeth meet in a scissors bite. His tail is docked.
4. Disqualifications: He should not stand over the height specification. A solid-colored dog should not have white markings anywhere except on his chest and throat. A parti-color or tricolor dog should not have 90% or more of one color. A black-and-tan dog should not be missing tan markings over his eyes, on his muzzle and cheeks, under his ears, on his feet and legs, or under his tail.

Health Problems

He is susceptible to PRA, cataracts, glaucoma, eyelid/eyelash/retinal abnormalities, skin conditions, hemophilia, ear infections, heart disease, and epilepsy. Buy only from CERF-registered parents.

Cautions when Buying

THERE ARE MANY COCKERS AROUND—BE VERY CAREFUL. Cockers are divided into field types (although there are few field Cockers today) and show types. Field Cockers have shorter coats for working in thickets. You shouldn't expect a Cocker from show lines to hunt, but either type usually makes a fine pet. If you buy this breed from a poor breeder, you'll probably end up with a flat-headed, skinny-muzzled, short-eared, sparse-coated Cocker with serious health and behavioral problems. Don't choose a timid or wildly excitable Cocker puppy.

English Cocker Spaniel

Fine for Novice Owners
Good with Children
Small in Size
Feathered Medium-Length Coat

	High	Med	Low
EXERCISE REQUIRED		✖	
TRIMMING/CLIPPING REQUIRED	✖		
AMOUNT OF SHEDDING		✖	
ACTIVITY INDOORS		✖	
EASE OF TRAINING	✖		
SOCIABILITY WITH STRANGERS	✖		

Temperament

He's lively, cheerful, gentle, and playful. He does fine in the city, but having more sporting instincts than his American cousin, he should have more exercise. He gets along well with strangers and other animals, but some are timid, so he should be accustomed to people and noises at an early age. He's eager to please and responds well to obedience training, but he's so sensitive and submissive that he should be trained persuasively, never sharply. Because he's less popular than the American Cocker, he has not been exploited as much by poor breeders and is thus a much safer bet as a good-tempered pet. He can be hard to housebreak.

History

He flushes birds from their hiding places and then retrieves them when the hunter shoots. He was developed in England, and his name comes from the woodcock game bird. He is a family hunting dog and companion. AKC popularity: 71st of 130 breeds.

Physical Features

1. He stands 15–17 inches and weighs 26–34 pounds.
2. His coat is silky, with longer hair on his chest, stomach, legs, and tail, and needs brushing and combing twice a week; scissoring and electric clipping every three months. Colors include blue roan (a speckled mixture of steel blue and white) and other roans, solid colors, black-and-tan (mostly black with tan points), parti-color (black-and-white, tan-and-white), and tricolor (black/tan/white).
3. His ears hang down; when pulled forward, they reach at least to his nose. His eyes are dark brown in dark-colored dogs, hazel in light-colored dogs. His nose is black on dark-colored dogs, brown on light-colored dogs. His teeth meet in a scissors or level bite. His tail is docked.
4. Faults: A dog of a solid color should not have white feet—and such a dog does *not* qualify as a parti-color just because he has two colors on him. Partis must have body markings well broken up.

Health Problems

He is susceptible to PRA, eyelid/eyelash abnormalities, and ear infections. Buy only from CERF-registered parents.

Cautions when Buying

English Cockers are divided into field types and show types. Field Cockers have shorter coats for working in thickets. You shouldn't expect an English Cocker from show lines to work like an English Cocker from field lines, but either type from a good breeder usually makes a fine pet. Don't choose a timid English Cocker puppy.

English Springer Spaniel

Fine for Novice Owners
Good with Children
Medium in Size
Feathered Medium-Length Coat

	High	Med	Low
EXERCISE REQUIRED		🦴	
TRIMMING/CLIPPING REQUIRED	🦴		
AMOUNT OF SHEDDING		🦴	
ACTIVITY INDOORS		🦴	
EASE OF TRAINING	🦴	🦴	
SOCIABILITY WITH STRANGERS	🦴	🦴	

Temperament

He is a cheerful, playful, high-spirited tail wagger. He's so adaptable that he does fine in the city if given plenty of attention and daily exercise, but some can be destructive and noisy if left alone too much. Some are more quiet, laid-back, and reserved than others, and some are timid, so he should be accustomed to people and noises at an early age. He gets along well with strangers and other animals. Although he can be boisterous, he's eager to please and responds well to obedience training that is not sharp or harsh. Some Springers develop odd phobias (an exaggerated fear of mustaches, floppy hats, fire hydrants, and so on).

History

He flushes birds from their hiding places and then retrieves them when the hunter shoots. He was developed in England, and his name comes from his duties of "springing" the pheasant into the air. He is a family hunting dog and companion. He is one of the most popular hunting dogs in the Northeast. AKC popularity: 22nd of 130 breeds.

Physical Features

1. He stands 19–20 inches and weighs 45–55 pounds.
2. His coat may be straight or wavy, with longer hair on his chest, stomach, legs, and tail, and needs brushing and combing twice a week. He needs scissoring and electric clipping every three months. He is black-and-white or liver- (brown) and-white. Tan may be added to form a tricolored dog. Blue roan (a speckled mixture of steel blue and white) and liver roan are less common.
3. His ears hang down. His eyes are brown in black dogs, hazel in liver dogs, with a kindly expression. His nose is black on black dogs, brown on liver dogs. His teeth meet in a scissors or level bite. His tail is docked.
4. Faults: He should not be red, orange, or lemon.

Health Problems

He is susceptible to hip dysplasia, PRA, eyelid/retinal abnormalities, ear infections, and skin conditions. Buy only from OFA- and CERF-registered parents.

Cautions when Buying

THERE ARE MANY SPRINGERS AROUND—BE VERY CAREFUL. Springers are divided into field types and show types. Field Springers are smaller, shorter-coated, and often have random patches and spots rather than the concentrated areas of color seen on show Springers. You shouldn't expect a Springer from show lines to hunt like a Springer from field lines, but either type from a good breeder usually makes a fine pet. If you buy this breed from a poor breeder, you could end up with a hyperactive, flaky Springer. Don't choose a wildly exuberant or excitable Springer puppy, and don't choose a timid puppy.

Field Spaniel

Fine for Novice Owners
Good with Children
Medium in Size
Feathered Medium-Length Coat

	High	Med	Low
EXERCISE REQUIRED		✦	
TRIMMING/CLIPPING REQUIRED		✦	
AMOUNT OF SHEDDING		✦	
ACTIVITY INDOORS	✦		
EASE OF TRAINING	✦		
SOCIABILITY WITH STRANGERS	✦		

Temperament

He is levelheaded, mild mannered, and reliable. He has more sporting instincts than most other spaniels and may be better in the suburbs or country, but he'd probably be okay in the city if given long daily walks and, as often as possible, taken for a run. He gets along well with strangers and other animals, but some are timid, so he should be accustomed to people and noises at an early age. This easygoing, willing dog responds well to obedience training that is not sharp or harsh.

History

Developed in England, he flushes birds from their hiding places and then retrieves them when the hunter shoots. The second-rarest of the nine spaniels, he deserves much greater popularity, but with the Cocker and Springer spaniels around, most people have never even heard of him. He is a family hunting dog and companion. AKC popularity: 125th of 130 breeds.

Physical Features

1. He stands about 18 inches and weighs 35–50 pounds.
2. His coat is silky, with longer hair on his chest, stomach, legs, and ears, and needs brushing and combing twice a week. Straggly hairs should be scissored every three months. He is solid black, solid liver (brown), golden liver, mahogany red, or roan (a speckled mixture of a dark color and white). He can also be one of these colors with tan markings over his eyes and on his cheeks and feet.
3. His ears hang down. His eyes are dark in black dogs, hazel in others, with a serious expression. His nose is black on black dogs, possibly brown on others. His teeth meet in a scissors or level bite. His tail is docked.
4. Faults: He should not be any color other than those listed above.

Health Problems

He is susceptible to hip dysplasia, retinal (eye) problems, and ear infections. Buy only from OFA-registered parents. At three to four months of age, most Field Spaniel puppies go through "collapsing front," an unusual condition where the front knees and ankles weaken and cause the puppy to flop about like a seal. During this trying time, the puppy needs rest, moderate exercise, and vitamin supplements. By six months of age, the puppy is strong again. Veterinarians who are unfamiliar with Field Spaniels often misdiagnose the condition as rickets.

Cautions when Buying

Don't choose the most independent Field Spaniel puppy, and don't choose a timid puppy.

Irish Water Spaniel

For Experienced Owners Only
Good with Older, Considerate Children
Large in Size
Curly Coat

	High	Med	Low
EXERCISE REQUIRED		●	
TRIMMING/CLIPPING REQUIRED		●	
AMOUNT OF SHEDDING			●
ACTIVITY INDOORS		●	
EASE OF TRAINING		●	
SOCIABILITY WITH STRANGERS			●

Temperament

He is unlike all other spaniels. This bold, confident, assertive dog has a distinct, purposeful presence. Many are loving family dogs, but many are one-person dogs. Because he needs plenty of daily exercise and loves the outdoors (including swimming), he does best in the suburbs or country. Reserved and protective with strangers, he should be accustomed to people and noises at an early age. He can be aggressive with other dogs. He's stubborn and independent but capable of learning a great deal. He will respond well to obedience training that is firm and consistent but not heavy-handed or too repetitious. Given early training and socialized well in a quiet household with access to space and exercise, this is a fun-to-be-with dog. Some are timid, nervous, or suspicious, and some are prone to snapping when startled or annoyed. He can refuse commands from family members who have not established leadership over him. Don't play aggressive games like wrestling or tug-of-war.

History

Despite the "spaniel" in his name, he usually retrieves ducks from the water. He can work like a spaniel, flushing birds from thickets, but his curly coat burrs up quickly. A very old breed from Ireland, he is a family hunting dog and companion. AKC popularity: 122nd of 130 breeds.

Physical Features

1. He stands 21–24 inches and weighs 45–65 pounds.
2. His coat is a mass of short, crisp ringlets, with longer curls on his ears and on the top of his head (topknot), but with a smooth face. His tail has some curls at the base, but otherwise it is a smooth "rat tail." He needs brushing thoroughly twice a week or his coat will be a mess, and he needs expert scissoring every three months. He is solid liver (brown).
3. His ears hang down; when pulled forward, they nearly reach his nose. His eyes are hazel with a quizzical expression. His nose is brown. His teeth meet in a level or scissors bite.
4. Faults: He should not have white hairs on his chest.

Health Problems

He is susceptible to hip dysplasia and hypothyroidism. Buy only from OFA-registered, thyroid-checked parents.

Cautions when Buying

Don't choose the boldest or most independent Irish Water Spaniel puppy or a timid or excitable puppy.

Sussex Spaniel

Fine for Novice Owners
Good with Children
Medium in Size
Feathered Medium-Length Coat

	High	Med	Low
EXERCISE REQUIRED		●	
TRIMMING/CLIPPING REQUIRED		●	
AMOUNT OF SHEDDING		●	
ACTIVITY INDOORS		●	
EASE OF TRAINING			●
SOCIABILITY WITH STRANGERS		●	

Temperament

He is slower, steadier, less playful, and less demonstrative than most other spaniels. He is a calm, deliberate friend, rather than an outgoing playmate. This heavyset field dog does better in the suburbs or country but should be okay in the city if exercised enough. He is usually friendly with strangers but can be aggressive with strange dogs. His stubbornness requires firm but patient obedience training. He could growl or snap if irritated or harshly disciplined. He howls a lot and can be hard to housebreak.

History

He flushes birds from their hiding places and then retrieves them when the hunter shoots. Developed in Sussex County, England, he is unique among the spaniels in that he bays when he hunts. The rarest of the nine spaniels, he deserves greater popularity, but with the Cocker and Springer spaniels around, most people have never even heard of him. He is primarily a companion, very occasionally a family hunting dog. AKC popularity: 128th of 130 breeds.

Physical Features

1. He stands 16–18 inches and weighs 35–45 pounds.
2. His coat is straight or wavy, with longer hair on his chest, stomach, legs, and tail, and needs brushing and combing twice a week. Straggly hairs should be scissored every three months. He is golden liver (golden brown) with yellow-tipped hairs.
3. His ears hang down. His eyes are hazel with a soft expression. His nose is brown. His teeth meet in a scissors or level bite. His tail is docked.
4. Faults: He should not have white hairs on his chest.

Health Problems

He is susceptible to ear infections.

Cautions when Buying

Don't choose the boldest, most independent Sussex puppy.

Welsh Springer Spaniel

Fine for Novice Owners
Good with Children
Medium in Size
Feathered Medium-Length Coat

	High	Med	Low
EXERCISE REQUIRED		🦴	
TRIMMING/CLIPPING REQUIRED		🦴	
AMOUNT OF SHEDDING		🦴	
ACTIVITY INDOORS		🦴	
EASE OF TRAINING	🦴	🦴	
SOCIABILITY WITH STRANGERS			🦴

Temperament

He is quiet, sensible, steady, less exuberant and less outgoing than the English Springer. A hardy, dependable worker in the field, he loves the outdoors but should do okay in the city if given long daily walks and, as often as possible, taken for a run. He's reserved with strangers and sometimes timid, so he should be accustomed to people and noises at an early age. He gets along very well with other animals. He's independent but also sensitive—he responds well to early, consistent obedience training that is not too harsh.

History

He flushes birds from their hiding places and then retrieves them when the hunter shoots. Developed in Wales, he is one of the oldest of the nine spaniel breeds. He is a family hunting dog and companion. AKC popularity: 98th of 130 breeds.

Physical Features

1. He stands 16–17 inches and weighs 38–45 pounds.
2. His coat is straight and silky, with longer hair on his chest, stomach, legs, and tail, and should be brushed and combed twice a week. Straggly hairs should be scissored every three months. He is red-and-white.
3. His ears hang down. His eyes are dark or hazel. His nose is dark or flesh colored. His teeth meet in a scissors or level bite. His tail is docked.

Health Problems

He is susceptible to PRA, ear infections, and epilepsy. Buy only from CERF-registered parents.

Cautions when Buying

Don't choose the most independent Welsh Springer puppy or a timid puppy.

Vizsla (VEESH-la)

For Experienced Owners Only
Good with Children
Large in Size
Smooth Coat

	High	Med	Low
EXERCISE REQUIRED	●		
TRIMMING/CLIPPING REQUIRED			●
AMOUNT OF SHEDDING		●	
ACTIVITY INDOORS		●	
EASE OF TRAINING		●	
SOCIABILITY WITH STRANGERS		●	

Temperament

He is gentle, good natured, and demonstrative with his own family. Agile and light on his feet, he needs enough running exercise that he belongs in the suburbs or country. Too much confinement makes him hyperactive and destructive—he is a notorious chewer. He gets along well with strangers and other animals, but some are timid, excitable, or easily startled, so he should be accustomed to people and noises at an early age. He should live with a quiet, calm (but athletic) family who will keep him relaxed and exercised. He's stubborn and easily distracted, but he does respond well to calm, firm, persistent training that does not include a lot of jerking around. Some can be hard to housebreak.

History

He freezes into a point when he scents a hiding bird and when his owner has shot it, he retrieves. A very old breed from the Hungarian plains, he roamed with the nomadic Magyar tribes, hunting partridge and rabbits. He is a family hunting dog and companion. AKC popularity: 59th of 130 breeds.

Physical Features

1. Males stand 22–24 inches, females stand 21–23 inches. He weighs 50–60 pounds.
2. His hard coat needs only a quick brushing once a week. He is solid golden rust, but small white spots on his chest or toes are allowed.
3. His ears hang down. His eyes and nose are light brown. His teeth meet in a scissors bite. His tail is docked.
4. Faults: His eyes should not be yellow. He should not have any black hairs in his coat.
5. Disqualifications: His nose should not be solid black. He should not have any white hairs, except for a few on his chest or toes. He should not stand more than 1½ inches over or under the height specification.

Health Problems

He is susceptible to hip dysplasia. Buy only from OFA-registered parents.

Cautions when Buying

Don't choose the most independent, excitable, or energetic Vizsla puppy, and don't choose a timid puppy.

Weimaraner (WY-mah-rah-ner)

For Experienced Owners Only
Good with Older, Considerate Children
Large in Size
Smooth Coat

	High	Med	Low
EXERCISE REQUIRED	✹		
TRIMMING/CLIPPING REQUIRED			✹
AMOUNT OF SHEDDING		✹	
ACTIVITY INDOORS	✹		
EASE OF TRAINING			✹
SOCIABILITY WITH STRANGERS		✹	

Temperament

He is so assertive, bold, energetic, and in need of space and running exercise that he is completely unsuited to an apartment. Too much confinement and too little companionship make him hyperactive and destructive—he's a notorious chewer. Reserved and protective with strangers, he should be accustomed to people and noises at an early age. He can be aggressive with other dogs and should be watched around small pets like cats and rabbits. He must have a fenced yard because he is a roamer with predatory instincts. This headstrong breed needs rigorous obedience training with a strong, firm hand. If indulged or not well trained he'll be far too rambunctious, refusing commands from family members who have not established leadership over him. Overall, an experienced owner who gives him exercise and training will find him a loyal, capable gentleman of great presence and character. A novice with little time and space will find him a boisterous, dominant bully, difficult to control. He can be hard to housebreak, and he can bark up a storm if confined outside in a kennel.

History

He freezes into a point when he scents a hiding bird and when his owner has shot it, he retrieves. He was developed in the German republic of Weimar, was owned only by the aristocracy, and originally hunted big game like bear and mountain lion. The "Gray Ghost" is a family hunting dog and companion. AKC popularity: 50th of 130 breeds.

Physical Features

1. Males stand 25–27 inches and weigh 70–90 pounds. Females stand 23–25 inches and weigh 55–75 pounds.
2. His hard coat needs only a quick brushing once a week. He is gray (from dark mouse to light silver). A small white marking on his chest is okay.
3. His ears hang down. His eyes are amber (brownish yellow), gray, or blue-gray. His nose is gray. His teeth meet in a scissors bite. His tail is docked to 6 inches. His feet are webbed to aid in swimming.
4. Faults: He should not have any white hairs, except on his chest. His eyes should not be any other color except the three allowed. The inside of his mouth should not be mottled black.
5. Disqualifications: He should not stand more than 1 inch over or under the height specification. His coat should not be long or blue or black in color.

Health Problems

He is susceptible to hip dysplasia, bloat, bleeding disorders, allergies, cysts, and tumors. Buy only from OFA-registered parents.

Cautions when Buying

BE CAREFUL. This is a powerful, independent breed, and if you buy from a poor breeder or raise the dog incorrectly, you could end up with an overly dominant, poor-tempered, or skittish Weimaraner. Don't choose the boldest, most rambunctious, or most independent Weimaraner puppy or a timid puppy.

Wirehaired Pointing Griffon (Griff-ON)

Fine for Novice Owners
Good with Children if Raised with Children
Medium to Large in Size
Wiry Coat

	High	Med	Low
EXERCISE REQUIRED	🦴		
TRIMMING/CLIPPING REQUIRED		🦴	
AMOUNT OF SHEDDING			🦴
ACTIVITY INDOORS	🦴		
EASE OF TRAINING		🦴	
SOCIABILITY WITH STRANGERS		🦴	

Temperament

He has a pleasant, gentle disposition, but he's also a rugged dog who loves to run outdoors. He belongs in the quiet country, for he tends to be timid of city sights and sounds. Too much confinement makes him restless and nervous, and in close quarters he may bark excessively at strange noises. He's reserved and often timid with strangers, so he should be accustomed to people and noises at an early age. Some are excitable, nervous, or easily startled—he should live with a quiet, calm (but athletic) family who will keep him relaxed and exercised. He gets along well with other animals. He's easily distracted, but quite responsive to calm obedience training. He can be hard to housebreak.

History

He freezes into a point when he scents a hiding bird, and when his owner has shot it, he retrieves. He was developed in Holland and France by a Dutchman named Korthals, and in Europe the breed is often called Korthals Griffon. *Griffon* means a wiry-coated dog. He is a family hunting dog and companion. AKC popularity: 118th of 130 breeds.

Physical Features

1. Males stand 21½–23½ inches, females stand 19½–21½ inches. He weighs 45–60 pounds.
2. His coat is harsh, with bushy eyebrows and a beard, and needs brushing and combing twice a week. Dead hairs should be stripped twice a year. He is a combination of steel gray, off-white, and chestnut red. Solid chestnut is less common.
3. His ears hang down. His eyes are light brown or yellow. His nose is brown. His teeth meet in a scissors or level bite. His tail is docked.
4. Faults: He should not be black.

Health Problems

He is susceptible to hip dysplasia. Buy only from OFA-registered parents.

Cautions when Buying

Don't choose the most independent Griffon puppy, and don't choose a timid puppy.

The Hound Group

Afghan Hound · Basenji · Basset Hound
13-Inch Beagle and 15-Inch Beagle
Black and Tan Coonhound · Bloodhound · Borzoi
Standard Longhaired Dachshund and Miniature Longhaired Dachshund
Standard Smooth Dachshund and Miniature Smooth Dachshund
Standard Wirehaired Dachshund and Miniature Wirehaired Dachshund
American Foxhound · English Foxhound · Greyhound · Harrier
Ibizan Hound · Irish Wolfhound · Norwegian Elkhound
Otterhound · Petit Basset Griffon Vendeen
Pharaoh Hound · Rhodesian Ridgeback
Feathered Saluki and Smooth Saluki
Scottish Deerhound · Whippet

Afghan Hound

For Experienced Owners Only
Good with Older, Considerate Children
Large in Size
Long Coat

	High	Med	Low
EXERCISE REQUIRED	●		
TRIMMING/CLIPPING REQUIRED			●
AMOUNT OF SHEDDING		●	
ACTIVITY INDOORS			●
EASE OF TRAINING			●
SOCIABILITY WITH STRANGERS			●

participating in the sport of lure coursing (chasing a trash bag pulled along a ground wire). AKC popularity: 57th of 130 breeds.

Temperament

Some Afghans are nervous, some dignified, some clownish. Quiet indoors, he seems to do okay in the city if given long daily walks and frequent runs. Don't let him off the leash except in a safe, enclosed area, for he is unbelievably fast and can be out of sight in seconds. His high hipbones make him one of the most agile of all breeds and one of the best fence jumpers. Reserved and often timid with strangers, he should be accustomed to people and noises at an early age. He gets along well with other dogs but should be watched around small pets like cats and rabbits. He needs obedience training to control his occasional bumptiousness, to build his confidence, and to enhance your ability to communicate with him, but you must be patient and persuasive, for he is sensitive to leash jerking. His stubbornness takes the form of resistance rather than wild misbehavior: He simply braces his legs and refuses to walk. Never hit this gentle breed. He can snap suddenly if frightened or startled, and he can be hard to housebreak. Many breeders feel that the male makes the better family pet. He should live with a quiet, calm (but athletic) family who will keep him relaxed and exercised.

History

Called a sighthound because he uses his keen sight to locate his prey, originally he ran down gazelles, hares, and leopards in the Afghanistan desert and mountains while the huntsmen followed on horseback. His forte is not straightaway speed, like a Greyhound, but incredible agility over rugged terrain. Today he is a companion, sometimes

Physical Features

1. Males stand 26–28 inches, females stand 24–26 inches. He weighs 50–60 pounds.
2. His coat is long and silky on the sides of his body, legs, ears, and the topknot on his head, but smooth on his face and back. Puppies develop this sharp contrast with maturity. He should be brushed and combed every other day or he will be a matted mess. Colors include black, cream, silver, brindle, and other solids and exotic patterns.
3. His ears hang down; when pulled forward, they nearly reach his nose. His eyes are dark with a far-off expression. His nose is black. His teeth meet in a level or scissors bite. His tail is fringed with hair and ringed or curved at the end.
4. Faults: He should not have white markings, especially on his head.

Health Problems

He is susceptible to hip dysplasia, cataracts, and paralysis. Because hip X-rays require anesthesia, many breeders do not check hips for dysplasia. This is true for all the sighthounds. He is sensitive to drugs (flea powders, anesthetics), so he should never be casually dosed with medication.

Cautions when Buying

Don't choose the most aloof or independent Afghan puppy or a timid puppy.

Basenji (Buh-SEN-jee)

For Experienced Owners Only
Good with Children if Raised with Children
Small in Size
Smooth Coat

	High	Med	Low
EXERCISE REQUIRED		✕	
TRIMMING/CLIPPING REQUIRED			✕
AMOUNT OF SHEDDING			✕
ACTIVITY INDOORS	✕		
EASE OF TRAINING			✕
SOCIABILITY WITH STRANGERS			✕

Temperament

He is high-spirited, playful, and curious, demanding to be in on everything. Physically unable to bark, he makes unusual chortling and yodeling sounds when happy. He's a clean dog, washing himself as a cat does. He does fine in the city, but he's active and needs regular exercise. Don't let him off the leash except in a safe, enclosed area, for he is a fast, agile, independent chaser and explorer who is difficult to catch. He can chew up your home if left alone too much, and he's an escape artist who can scale fences (and sometimes trees). You must stay one step ahead of this clever little dog. He is reserved with strangers and can be scrappy with other animals. He's stubborn, but does respond to the firm obedience training necessary to keep him well mannered and controlled. Don't play aggressive games like wrestling or tug-of-war.

History

Once called the Congo Dog, he dashed through African woods driving small game into the tribal huntsmen's nets. Because he could not bark, he wore wooden rattles or a bell around his neck to indicate his whereabouts. Today he is a companion, sometimes participating in the sport of lure coursing (chasing a trash bag pulled along a ground wire). AKC popularity: 60th of 130 breeds.

Physical Features

1. He stands 16–17 inches and weighs 20–25 pounds.
2. His coat is soft and loose skinned and needs only a quick brushing once a week. He is chestnut red, black-and-tan, or black. Also, brindle (brownish with black stripes and flecks) has just been accepted as an allowable color. He always has white feet, chest, and tail tip; white legs, face blaze, and neck collar are optional.
3. His ears prick up. His eyes are dark hazel. His nose is black, sometimes with a pink tinge. His teeth meet in a scissors bite. His tail curls tightly to either side of his back.
4. Faults: He should not be any color (such as cream) other than those listed.

Health Problems

He is susceptible to hernias, fanconi (severe kidney disorder), some eye problems, and digestive upsets. Females usually come into heat only once a year instead of twice, which is normal for other domestic breeds.

Cautions when Buying

Don't choose the boldest, most energetic, or most independent Basenji puppy, or a timid puppy.

Basset Hound

Fine for Novice Owners
Good with Children
Medium in Size
Smooth Coat

	High	Med	Low
EXERCISE REQUIRED		✦	
TRIMMING/CLIPPING REQUIRED			✦
AMOUNT OF SHEDDING		✦	
ACTIVITY INDOORS			✦
EASE OF TRAINING		✦	
SOCIABILITY WITH STRANGERS		✦	

Temperament

He is among the mildest mannered of all breeds. Some are dignified, some are clownish, but almost all are good-natured, peaceful, and reliable. He's fine in the city, but stronger and heavier than one might think. He needs daily exercise to stay fit, even if he doesn't appear to want it. Lazy owners have fat Bassets with health problems. Most are friendly with strangers, some are slightly reserved. He gets along well with other animals. He's a bit stubborn but responds well to patient obedience training that includes much praise and little jerking around. This sweet breed should never be harshly disciplined. He does not always obey commands quickly and he can exhibit an amusing sense of humor while being disobedient, but he's seldom a real problem. He can be hard to housebreak, and he bays a lot in a deep, soulful voice.

History

A French scenthound descended from the Bloodhound, he trails rabbits, baying as he goes. His name comes from the French *bas*, meaning "low to the ground." There are several varieties of Basset in Europe, some taller and with less-crooked legs than the AKC version we are accustomed to. He is primarily a companion, very occasionally a family hunting dog. AKC popularity: 18th of 130 breeds.

Physical Features

1. He stands up to 15 inches and weighs 40–55 pounds.
2. His hard coat is loose skinned and needs only a quick brushing once a week. He is a combination of black, white, and/or tan.
3. His ears hang down; when pulled forward, they reach past his nose. His eyes are brown with a sad expression. His nose is black or brown. His teeth meet in a scissors or level bite.
4. Disqualifications: His coat should not be long. He should not stand over the height specification. His knees should not knuckle forward.

Health Problems

He is susceptible to bloat, spinal disc problems, ear infections, glaucoma, and eyelid abnormalities.

Cautions when Buying

THERE ARE MANY BASSETS AROUND—BE VERY CAREFUL. Bassets are divided into field types and show types. Field Bassets are taller, leggier, faster, and less loose skinned than show Bassets. You shouldn't expect a Basset from show lines to hunt like a Basset from field lines, but either type from a good breeder usually makes a fine pet. If you buy this breed from a poor breeder, you could wind up with a sickly, shy, noisy, neurotic Basset. Don't choose a timid Basset puppy.

13-Inch Beagle and 15-Inch Beagle

These are varieties of the same breed.

Fine for Novice Owners
Good with Children
Small in Size
Smooth Coat

	High	Med	Low
EXERCISE REQUIRED		●	
TRIMMING/CLIPPING REQUIRED			●
AMOUNT OF SHEDDING		●	
ACTIVITY INDOORS	●		
EASE OF TRAINING		●	
SOCIABILITY WITH STRANGERS	●		

Temperament

He is good-natured, cheerful, and gentle. He's so adaptable that he does fine in the city if given daily walks and occasionally taken for a run. He is a sniffer and an explorer—don't let him off the leash except in a safe, enclosed area or he will fasten his nose to the ground and take off. He gets along well with strangers and other animals, but some are timid, so he should be accustomed to people and noises at an early age. His stubbornness calls for early, firm obedience training, but he is highly sensitive, so use much praise and little jerking around. He may snap defensively if harshly disciplined—no Beagle should ever be hit. To come when called is the most important command to be learned by this curious little hound. He barks and howls a lot, especially if left alone, and he can be hard to housebreak. Some Beagles develop odd phobias (an exaggerated fear of beards, eyeglasses, mailboxes, and so on).

History

A scenthound from England, he trails rabbits either singly or in packs, baying as he goes. His name comes from a group of old hunting hounds called *begles*. He is a family hunting dog and companion. AKC popularity: 8th of 130 breeds.

Physical Features

1. He comes in two size varieties: A 13-inch Beagle stands 10–13 inches and weighs 18–20 pounds; a 15-inch Beagle stands 13–15 inches and weighs 20–30 pounds.
2. His hard coat needs only a quick brushing once a week. He is a combination of black, white, and/or tan.
3. His ears hang down; when pulled forward, they nearly reach his nose. His eyes are brown or hazel with a pleading expression. His nose is black. His teeth meet in a level bite.
4. Disqualifications: He should not stand more than 15 inches high.

Health Problems

He is susceptible to heart disease, spinal disc problems, bleeding disorders, epilepsy, skin conditions, glaucoma, and cataracts.

Cautions when Buying

THERE ARE MANY BEAGLES AROUND—BE VERY CAREFUL. Beagles are divided into field types and show types. A Beagle from show lines is more compactly built, and you shouldn't expect him to hunt like a Beagle from field lines. The Beagles that make the best pets tend to come from the show lines, because some field Beagles can be high-strung. If you buy this breed from a poor breeder, you could wind up with a shy, hyperactive, noisy, snappish, sickly Beagle. Don't choose the most independent Beagle puppy, and don't choose a timid or wildly excitable puppy. *Caution:* Some questionable breeders produce "Pocket Beagles," tiny creatures similar to "Teacup Poodles" and prone to the same bizarre physical and behavioral problems.

Black and Tan Coonhound

For Experienced Owners Only
Good with Children
Large in Size
Smooth Coat

	High	Med	Low
EXERCISE REQUIRED	✦		
TRIMMING/CLIPPING REQUIRED			✦
AMOUNT OF SHEDDING		✦	
ACTIVITY INDOORS			✦
EASE OF TRAINING			✦
SOCIABILITY WITH STRANGERS		✦	

Temperament

He is good-natured and easygoing, but he's a big, powerful, hardy hound, so energetic and in need of running exercise and space that he belongs in the suburbs or country with an athletic owner who loves the outdoors. He is a sniffer and an explorer, so he should not be let off the leash except in a safe, enclosed area, for he will drop his nose to the ground and be off. He's reserved with strangers, and some are protective. He can be aggressive with strange dogs. His strength and stubbornness call for early, firm, persistent obedience training—the commands of *heel*, *down*, and *come* are especially important in controlling this big breed. Some bay and howl a lot, especially if left alone too much.

History

Descended from British scenthounds, he was fully developed in the American Deep South along with other coonhound breeds (Redbone, Bluetick, Plott) that are recognized by the UKC but not the AKC. The Black and Tan may also be registered with the UKC, and many B/T owners register their dogs with both organizations or only with the UKC, because that organization concentrates on the field (hunting) trials at which this breed excels. He is a hunting dog who bays upon treeing the coon, and also a companion dog. AKC popularity: 106th of 130 breeds.

Physical Features

1. Males stand 25–27 inches, females stand 23–25 inches. He weighs 75–90 pounds.

2. His hard coat is loose skinned and needs only a quick brushing once a week. He is black, with tan markings above his eyes, on his muzzle, chest, legs, and as streaks on his toes.
3. His ears hang down; when pulled forward, they reach beyond his nose. His eyes are dark brown or hazel. His nose is black. His teeth meet in a scissors bite.
4. Faults: His eyes should not be yellow.
5. Disqualifications: He should not have any white spots that are more than 1 inch in diameter.

Health Problems

He is susceptible to hip dysplasia.

Cautions when Buying

Black and Tans are divided into field types and show types. A Black and Tan from show lines is bigger and longer eared. The Black and Tans that make the best pets tend to come from the show lines, because field Coonhounds can be too energetic and too businesslike to be at their best as pets. Don't choose the most rambunctious or independent Black and Tan puppy.

Bloodhound

For Experienced Owners Only
Good with Children
Large to Giant in Size
Smooth Coat

	High	Med	Low
EXERCISE REQUIRED	●		
TRIMMING/CLIPPING REQUIRED			●
AMOUNT OF SHEDDING		●	
ACTIVITY INDOORS			●
EASE OF TRAINING			●
SOCIABILITY WITH STRANGERS	●		

Temperament

He is patient, easygoing, and mild mannered, but so enthusiastic and rambunctious outdoors that he belongs in the suburbs or country with an athletic owner who loves the outdoors. He is a sniffer and an explorer, so he should not be let off the leash except in a safe, enclosed area, for once his nose locks onto a fascinating scent, it is almost impossible to regain his attention. Good-natured and gregarious, he likes the company of people and other animals, but some males can be aggressive with other male dogs. His great strength and stubbornness call for a strong, firm owner, but this kindly breed is quite sensitive and should never be harshly disciplined or hit. The commands of *heel, down,* and *come* are especially important in controlling him. He bays a lot, and he drools.

History

This ancient scenthound originated in the Mediterranean area, near Rome. The monks at St. Hubert's Monastery kept the breed "of pure blood of the highest order" and only aristocratic "bluebloods" owned him, hence the name "blooded hound." Whether trailing lost children, disoriented old people, or vicious criminals, this face kisser only tracks—never attacks. He is almost impossible to shake off a track, and documented tracks 100 miles long and 100 hours old have made his olfactory evidence admissible in a court of law. He is a companion and can be so valuable in a community emergency that all Bloodhound owners should teach their dogs to track. AKC popularity: 62nd of 130 breeds.

Physical Features

1. Males stand 25–27 inches, females stand 23–25 inches. He weighs 80–110 pounds.
2. His hard coat is loose skinned and needs only a quick brush once a week. He is black-and-tan, red-and-tan, or solid tan. He can be flecked with white or have a little white on his chest, feet, and tail tip. Puppies are born dark and unwrinkled.
3. His ears hang down; when pulled forward, they reach beyond his nose. His eyes are hazel or yellow with a wise expression. His teeth meet in a scissors or level bite.

Health Problems

He is susceptible to hip dysplasia, bloat, and eyelid abnormalities. Buy only from OFA-registered parents.

Cautions when Buying

Don't choose the boldest, most rambunctious, or most independent Bloodhound puppy, and don't choose a timid puppy.

Borzoi (BOR-zoy)

For Experienced Owners Only
Good with Older, Considerate Children
Large to Giant in Size
Feathered Medium-Length Coat

	High	*Med*	*Low*
EXERCISE REQUIRED		✖	
TRIMMING/CLIPPING REQUIRED			✖
AMOUNT OF SHEDDING		✖	
ACTIVITY INDOORS			✖
EASE OF TRAINING			✖
SOCIABILITY WITH STRANGERS			✖

Temperament

He is quiet, independent, not very demonstrative, and sometimes compared to a cat in temperament. Although often seen in the city, he's really best in the suburbs or country where he has space to move. However, he should not be let off the leash except in a safe, enclosed area, for he is a chaser and his floating gallop is unbelievably fast. He is reserved with strangers, so he should be accustomed to people at an early age. He can be aggressive with strange dogs and should be watched around small pets like cats and rabbits. He's not recommended around mischievous children because, although usually sweet and docile, he can suddenly snap with lightning reflexes if startled, teased, or touched unexpectedly. Because he is highly sensitive, he should be trained with much more praise than correction, and never jerked around or hit. He should live with a quiet, calm (but athletic) family who will keep him relaxed and exercised.

History

Called a sighthound because he uses his keen sight to locate prey, he comes from Czarist Russia, where he was owned by the aristocracy and hunted in packs. He ran down wolves and other prey across open terrain while the huntsmen followed on horseback. He has been used in America to run down coyotes, but he is primarily a companion, sometimes participating in the sport of lure coursing (chasing a trash bag pulled along a ground wire). AKC popularity: 72nd of 130 breeds.

Physical Features

1. Males stand at least 28 inches and weigh 75–105 pounds. Females stand at least 26 inches and weigh 55–90 pounds.
2. His coat is silky, with longer hair on his chest, stomach, legs, and tail, but with very short hair on his face. His coat should be brushed and combed twice a week. He's usually white, with markings of lemon, tan, gray, or black, but any color or color combination is acceptable.
3. His ears fold back against his head but prick up halfway when he's excited. His eyes are dark with a soft expression. His nose is black. His teeth meet in a scissors or level bite.

Health Problems

He is susceptible to bloat. He is sensitive to drugs (anesthetics, flea powders), so he should never be casually dosed with medication.

Cautions when Buying

BE CAREFUL. This is a fast, agile, powerful, independent breed, and if you buy from a poor breeder or raise the dog incorrectly, you could wind up with an unpredictable Borzoi. Don't choose the most aloof, withdrawn, independent Borzoi puppy or a timid puppy.

Standard Longhaired Dachshund (DAHKS-hund) and Miniature Longhaired Dachshund

There are six varieties of Dachshund.

Fine for Novice Owners
Good with Older, Considerate Children
Little to Small in Size
Feathered Medium-Length Coat

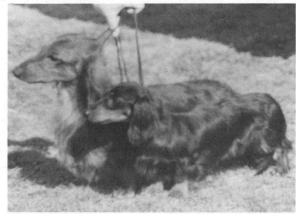

	High	Med	Low
EXERCISE REQUIRED			✦
TRIMMING/CLIPPING REQUIRED			✦
AMOUNT OF SHEDDING		✦	
ACTIVITY INDOORS	✦		
EASE OF TRAINING		✦	
SOCIABILITY WITH STRANGERS			✦

Temperament

He is curious and playful, bold and brash, similar to a terrier in his demands to be in on everything. He thrives on companionship and is great in the city but does like his walks. He's adaptable and makes an excellent traveling companion. He's reserved with strangers, usually fine with other animals. This clever, obstinate little dog is responsive to firm, patient, persuasive training, but he occasionally disobeys with a comical sense of humor. He should never be harshly disciplined or hit, for he could snap. Don't play aggressive games like wrestling or tug-of-war. He barks, digs, and can be hard to housebreak. The Longhaired variety is considered by some enthusiasts to be quieter and more docile than the Smooth and Wirehaired varieties.

History

Incredibly, he was once a thirty-five-pound hunter of the fierce German badger, and his name comes from *dachs* (badger) *hund* (dog). He was bred down in size to hunt rabbits, and he still does so in Europe. In America he is a companion. AKC popularity: 10th of 130 breeds.

Physical Features

1. He comes in two size varieties: Standard, about 9 inches and 10–20 pounds; Miniature, about 5 inches and under 10 pounds.
2. His coat is silky, with longer hair on his chest, stomach, legs, and tail, and needs to be brushed and combed twice a week. He is usually black-and-tan or solid red (actually brownish). He can also be white-and-tan, dappled (a swirling mixture of brown, gray, white, and black), or other colors.
3. His ears hang down. His eyes are brown, but bluish white eyes are okay in dappled dogs. His nose is black, brown, or gray. His teeth meet in a scissors bite.
4. Faults: His knees should not knuckle over. His toes should not turn inward. His back should not be sunken like that of a swaybacked horse.

Health Problems

He is a long-lived breed (fifteen years) but susceptible to spinal disc problems (don't let him jump off furniture), diabetes, urinary stones, eye disorders, skin conditions, and heart disease.

Cautions when Buying

THERE ARE MANY DACHSHUNDS AROUND— BE VERY CAREFUL. If you buy this breed from a poor breeder you could wind up with a noisy, sickly, snappish Dachshund. Don't choose the boldest, most scrappy, or most independent Dachshund puppy, and don't choose a timid puppy.

Standard Smooth Dachshund (DAHKS-hund) and Miniature Smooth Dachshund

There are six varieties of Dachshund.

Fine for Novice Owners
Good with Older, Considerate Children
Little to Small in Size
Smooth Coat

	High	Med	Low
EXERCISE REQUIRED			⬮
TRIMMING/CLIPPING REQUIRED			⬮
AMOUNT OF SHEDDING		⬮	
ACTIVITY INDOORS	⬮		
EASE OF TRAINING		⬮	
SOCIABILITY WITH STRANGERS			⬮

Temperament

He is curious and playful, bold and brash, similar to a terrier in his demand to be in on everything. He thrives on companionship and is great in the city but does like his walks. He's adaptable and makes an excellent traveling companion. He's reserved with strangers, usually fine with other animals. This clever, obstinate little dog is responsive to firm, patient, persuasive training, but he occasionally disobeys with a comical sense of humor. He should never be harshly disciplined or hit, for he could snap. Don't play aggressive games like wrestling or tug-of-war. He barks, digs, and can be hard to housebreak. The Smooth and Wirehaired varieties are considered by some enthusiasts to be bolder and more active than the Longhaired variety.

History

Incredibly, he was once a thirty-five-pound hunter of the fierce German badger, and his name comes from *dachs* (badger) *hund* (dog). He was bred down in size to hunt rabbits, and he still does so in Europe. In America he is a companion. AKC popularity: 10th of 130 breeds.

Physical Features

1. He comes in two size varieties: Standard, about 9 inches and 10–20 pounds; Miniature, about 5 inches and under 10 pounds.
2. His hard coat needs only a quick brush once a week.

He is usually black-and-tan or solid red (actually brownish). He can also be white-and-tan, dappled (a swirling mixture of brown, gray, white, and black), or other colors.
3. His ears hang down. His eyes are brown, but bluish white eyes are okay in dappled dogs. His nose is black, brown, or gray. His teeth meet in a scissors bite.
4. Faults: His knees should not knuckle over. His toes should not turn inward. His back should not be sunken like that of a swaybacked horse.

Health Problems

He is a long-lived breed (fifteen years), but susceptible to spinal disc problems (don't let him jump off furniture), diabetes, urinary stones, eye disorders, skin conditions, and heart disease.

Cautions when Buying

THERE ARE MANY DACHSHUNDS AROUND—BE VERY CAREFUL. If you buy this breed from a poor breeder you could wind up with a noisy, sickly, snappish Dachshund. Don't choose the boldest, most scrappy, or most independent Dachshund puppy, and don't choose a timid puppy.

Standard Wirehaired Dachshund (DAHKS-hund) and Miniature Wirehaired Dachshund

There are six varieties of Dachshund.

Fine for Novice Owners
Good with Older, Considerate Children
Little to Small in Size
Wiry Coat

	High	Med	Low
EXERCISE REQUIRED			●
TRIMMING/CLIPPING REQUIRED		●	
AMOUNT OF SHEDDING			●
ACTIVITY INDOORS	●		
EASE OF TRAINING		●	
SOCIABILITY WITH STRANGERS			●

Temperament

He is curious and playful, bold and brash, similar to a terrier in his demand to be in on everything. He thrives on companionship and is great in the city but does like his walks. He's adaptable and makes an excellent traveling companion. He's reserved with strangers, usually fine with other animals. This clever, obstinate little dog is responsive to firm, patient, persuasive training, but he occasionally disobeys with a comical sense of humor. He should never be harshly disciplined or hit, for he could snap. He barks, digs, and can be hard to housebreak. Don't play aggressive games like wrestling or tug-of-war. The Smooth and Wirehaired varieties are considered by some enthusiasts to be bolder and more active than the Longhaired variety.

History

Incredibly, he was once a thirty-five-pound hunter of the fierce German badger, and his name comes from *dachs* (badger) *hund* (dog). He was bred down in size to hunt rabbits, and he still does so in Europe. In America he is a companion. AKC popularity: 10th of 130 breeds.

Physical Features

1. He comes in two size varieties: Standard, about 9 inches and 10–20 pounds; Miniature, about 5 inches and under 10 pounds.
2. His coat is rough, with bushy eyebrows and a beard, and needs to be brushed and combed twice a week. Straggly hairs should be scissored and dead hairs stripped twice a year. He is usually black-and-tan or solid red (actually brownish). He can also be white-and-tan, dappled (a swirling mixture of brown, gray, white, and black), or other colors.
3. His ears hang down. His eyes are brown, but bluish white eyes are okay in dappled dogs. His nose is black, brown, or gray. His teeth meet in a scissors bite.
4. Faults: His knees should not knuckle over. His toes should not turn inward. His back should not be sunken like that of a swaybacked horse.

Health Problems

He is a long-lived breed (fifteen years) but susceptible to spinal disc problems (don't let him jump off furniture), diabetes, urinary stones, eye disorders, skin conditions, and heart disease.

Cautions when Buying

THERE ARE MANY DACHSHUNDS AROUND— BE VERY CAREFUL. If you buy this breed from a poor breeder, you could wind up with a noisy, sickly, snappish Dachshund. Don't choose the boldest, most scrappy, or most independent Dachshund puppy, and don't choose a timid puppy.

American Foxhound

For Experienced Owners Only
Good with Children
Large in Size
Smooth Coat

	High	Med	Low
EXERCISE REQUIRED	✦		
TRIMMING/CLIPPING REQUIRED			✦
AMOUNT OF SHEDDING		✦	
ACTIVITY INDOORS	✦		
EASE OF TRAINING			✦
SOCIABILITY WITH STRANGERS		✦	

Temperament

He is stable and good-natured. Some make good house pets, while others seem to prefer living kenneled outdoors with other hunting hounds. This energetic dog needs so much running exercise that he belongs in the suburbs or country with an athletic owner who loves the outdoors. Too much confinement makes him restless, rambunctious, and prone to destructive chewing. He is a sniffer and an explorer, so don't let him off the leash except in a safe, enclosed area, for he will fasten his nose to the ground and take off. Some are friendly with strangers, some are reserved, some are protective. He gets along well with other animals. Training takes a while, for he's stubborn and slow to obey; above all, he needs to learn to come when called. He bays a lot and can be hard to housebreak.

History

He is descended from British scenthounds but was developed mostly in Virginia. He usually hunts in packs, trailing foxes while the huntsmen follow on horseback. He is a more individualistic hunter than his cousin, the English Foxhound, as evidenced by his habit of baying only when he is leading the pack. He is primarily a hunting dog, occasionally a companion. AKC popularity: 127th of 130 breeds.

Physical Features

1. He stands 23–27 inches and weighs 60–80 pounds.
2. His hard coat needs only a quick brushing once a week. He is a combination of black, white, and tan.

3. His ears hang down. His eyes are brown or hazel, with a gentle expression. His teeth meet in a scissors or level bite.

Health Problems

He is a healthy breed, prone to no real problems.

Cautions when Buying

Foxhounds are divided into field types and show types, and you shouldn't expect a Foxhound from show lines to hunt like a Foxhound from field lines. The Foxhounds that make the best pets tend to come from the show lines, because field Foxhounds can be too energetic and businesslike to be at their best as a pet. Don't choose the most active or independent Foxhound puppy.

English Foxhound

For Experienced Owners Only
Good with Children
Large in Size
Smooth Coat

	High	Med	Low
EXERCISE REQUIRED	🦴		
TRIMMING/CLIPPING REQUIRED			🦴
AMOUNT OF SHEDDING		🦴	
ACTIVITY INDOORS	🦴		
EASE OF TRAINING			🦴
SOCIABILITY WITH STRANGERS		🦴	

Temperament

He is stable and good-natured. Some make good house pets, while others seem to prefer living kenneled outdoors with other hunting hounds. This energetic dog needs so much running exercise that he belongs in the suburbs or country with an athletic owner who loves the outdoors. Too much confinement makes him restless, rambunctious, and prone to destructive chewing. He is a sniffer and an explorer, so don't let him off the leash except in a safe, enclosed area, for he will fasten his nose to the ground and take off. Some are friendly with strangers, some are reserved. He gets along well with other animals. Training takes a while, for he's stubborn and slow to obey; above all, he needs to learn to come when called. He bays a lot and can be hard to housebreak.

History

Descended from British scenthounds, he usually hunts in packs, trailing foxes and baying while the huntsmen follow on horseback. He is primarily a hunting dog, occasionally a companion. AKC popularity: 130th of 130 breeds.

Physical Features

1. He stands 21–25 inches and weighs 60–70 pounds.
2. His hard coat needs only a quick brushing once a week. He is a combination of black, white, tan, and sometimes yellow.
3. His ears hang down; sometimes one inch is surgically cut off the tips to "round off" the ears. His eyes are brown or hazel, with a gentle expression. His teeth meet in a level bite.
4. Disqualifications: His teeth should not meet in an overshot or undershot bite.

Health Problems

He is a healthy breed, prone to no real problems.

Cautions when Buying

Foxhounds are divided into field types and show types, and you shouldn't expect a Foxhound from show lines to hunt like a Foxhound from field lines. The Foxhounds that make the best pets tend to come from the show lines, because field Foxhounds can be too energetic and businesslike to be at their best as a pet. Don't choose the most active or independent Foxhound puppy.

Greyhound

For Experienced Owners Only
Good with Older, Considerate Children
Large in Size
Smooth Coat

	High	Med	Low
EXERCISE REQUIRED	🦴		
TRIMMING/CLIPPING REQUIRED			🦴
AMOUNT OF SHEDDING		🦴	🦴
ACTIVITY INDOORS			🦴
EASE OF TRAINING		🦴	
SOCIABILITY WITH STRANGERS		🦴	

Temperament

He is quiet, clean, and gentle, but he belongs in the suburbs or country where he can get running exercise and won't be frightened by city sights and sounds. He should not be let off the leash except in a safe, enclosed area, for this agile chaser is the fastest dog in the world. Reserved and often timid with strangers, he should be carefully accustomed to people and noises at an early age. He's fine with other dogs, but should be watched around small pets like cats and rabbits. Training takes time and effort, for although he's not stubborn, he is nervous, sensitive, and easily distracted. You must be gentle and patient, not jerking him around, not scolding too much, and never hitting him. Obedience training is wonderful for building confidence in this sweet dog, and you will be better able to communicate with him. He can bolt if startled or suddenly touched. He can be hard to housebreak. He should live with a quiet, calm (but athletic) family who will keep him relaxed and exercised.

History

This dog was developed in ancient Egypt and called a sighthound because he uses his keen sight to locate prey and run it down across open terrain. He is used for racing— a bloody industry in which dogs chase a mechanical rabbit after training on live rabbits; ex-racers are usually destroyed. He is a companion, sometimes participating in the sport of lure coursing (chasing a trash bag pulled along a ground wire). AKC popularity: 121st of 130 breeds.

Physical Features

1. He stands 26–30 inches and weighs 60–90 pounds.
2. His hard coat needs only a quick brushing once a week. Colors include white, gray, fawn, patched, brindle (brownish with black stripes and flecks), and others.
3. His ears fold back against his head but prick up halfway when he is excited. His eyes are dark with a spirited expression. His nose is dark. His teeth meet in a level bite.

Health Problems

He is susceptible to bloat. He is sensitive to drugs (anesthetics, flea powders), so he should never be casually dosed with medication. He should be protected from extreme cold and heat.

Cautions when Buying

BE CAREFUL. Greyhounds are divided into racing types bred for speed and show types bred for good conformation and good temperament. The Greyhounds that make the best pets come from the show lines. Track Greyhounds are often unsocialized and high-strung to the point of being unreliable around people and other animals. However, Greyhound Rescue Leagues select the gentlest, most sweet-tempered ex-racers for placement into pet homes; this is a worthy endeavor and reports of success rates are extremely high. Don't choose the most independent Greyhound puppy, and don't choose a timid puppy.

Harrier

Fine for Novice Owners
Good with Children
Medium in Size
Smooth Coat

	High	Med	Low
EXERCISE REQUIRED	●		
TRIMMING/CLIPPING REQUIRED			●
AMOUNT OF SHEDDING		●	
ACTIVITY INDOORS	●		
EASE OF TRAINING		●	
SOCIABILITY WITH STRANGERS	●		

Temperament

He is stable and good-natured but energetic. He needs so much running exercise that he belongs in the suburbs or country with an athletic owner who loves the outdoors. Too much confinement makes him restless and rambunctious. He is a sniffer and explorer, so he should not be let off the leash except in a safe, enclosed area, for he cannot resist exciting scents or sudden movements. Most are friendly with strangers. He gets along well with other animals. Training takes a while, for he's a bit stubborn. Above all, he needs to learn to come when called. He bays a lot and can be hard to housebreak. Midway in size between a Beagle and a Foxhound, he makes a pleasant, adaptable pet when given enough exercise and training.

History

A scenthound from England, he trails hares (usually in packs) while the huntsmen follow on horseback. He is too fast for cottontails, ending the chase by sending them almost immediately into a hole, but he is just right for the larger hare, which will run all day and seldom hole up. Since there are few hares in America, he is more popular in Europe. His name comes from his quarry (the hare), or from the Continental word *harier*, meaning "hound." He is a hunting dog and companion. This breed should be much more popular than it is. AKC popularity: 129th of 130 breeds.

Physical Features

1. He stands 19–21 inches and weighs 40–55 pounds.
2. His hard coat needs only a quick brushing once a week. He is a combination of black, white, and tan; occasionally he is a mottled blue-gray-black, or a badger color.
3. His ears hang down. His eyes are brown or hazel. His nose is black. His teeth meet in a scissors bite.

Health Problems

He is a healthy breed, prone to no real problems.

Cautions when Buying

Harriers are divided into field types and show types. The Harriers that make the best pets tend to come from the show lines, because field Harriers can be too energetic and too businesslike to be at their best as pets. Don't choose the most energetic or independent Harrier puppy.

Ibizan Hound (Ih-BEEZ-an)

Fine for Novice Owners
Good with Children
Medium to Large in Size
Smooth Coat

	High	Med	Low
EXERCISE REQUIRED	▶◀		
TRIMMING/CLIPPING REQUIRED			▶◀
AMOUNT OF SHEDDING		▶◀	
ACTIVITY INDOORS			▶◀
EASE OF TRAINING		▶◀	
SOCIABILITY WITH STRANGERS		▶◀	

Temperament

He is a clean dog, quiet and well mannered indoors. He can adapt to the city, but he needs running exercise and should not be left alone too much. He's playful, athletic, light on his feet, and an excellent fence-jumper. This fast, agile dog should not be let off the leash except in a safe, enclosed area. He is sensibly friendly with strangers, but he knows his property and no one should just walk into an Ibizan's house without having first been welcomed by his owner. Some are timid, so he should be accustomed to people and noises at an early age. Most are good with other animals, although he should be watched around small pets like cats and rabbits. He enjoys obedience training, but he is easily bored by repetition and does have a mind of his own. He should be handled calmly and persuasively, for he is sensitive to correction.

History

Called a sighthound because he uses his keen sight to locate prey, he is an ancient Egyptian breed that was used to run down rabbits. When a statue of the canine god Anubis was uncovered in King Tut's tomb, the statue was believed to be that of an Ibizan, but the breed did not come by its name until it became popular on the Mediterranean island of Ibiza. Today he is a companion. AKC popularity: 117th of 130 breeds.

Physical Features

1. He stands 22½–27½ inches and weighs 42–50 pounds.
2. His coat is usually smooth (occasionally wiry) and needs only a quick brushing once a week. He is red (most common) or lion colored (tawny tan), with white feet, tail tip, chest, muzzle, and blaze. Solid red or solid white are uncommon.
3. His ears prick up. His eyes are light brown or amber (brownish yellow). His nose is flesh colored. His teeth meet in a scissors bite.

Health Problems

He is sensitive to drugs (anesthetics, flea powders), so he should never be casually dosed with medication.

Cautions when Buying

Don't choose the most independent Ibizan puppy or a timid puppy.

Irish Wolfhound

Fine for Novice Owners
Good with Children
Giant in Size
Wiry Coat

	High	Med	Low
EXERCISE REQUIRED		●	
TRIMMING/CLIPPING REQUIRED		●	
AMOUNT OF SHEDDING		●	
ACTIVITY INDOORS			●
EASE OF TRAINING		●	
SOCIABILITY WITH STRANGERS		●	

Temperament

This gentle giant is sometimes calm and dignified, sometimes playful, always easygoing and reliable. Although often seen in the city, he is so big that he does best in a suburban or country home where he has room to stretch out. He shouldn't be crammed into tight living quarters just because he good-naturedly accepts it, and he needs regular exercise to stay fit, whether he seems to want it or not. He is sensible with strangers, occasionally timid, good with other animals. A willing dog, he responds well to patient obedience training. You must be firm if he gets too rambunctious, but you should never be harsh with this sweet dog. Give him time to respond—he cannot be as quick and precise as a Toy Poodle. His only real behavioral problem is clumsiness—he may be too big and awkward for toddlers. This breed takes several years to mature; a two-year-old Wolfhound is still a puppy.

History

This is the tallest breed in the world and it was developed in feudal Ireland for hunting wolves and six-foot-tall elk. Today he is a companion, sometimes participating in the sport of lure coursing (chasing a trash bag pulled along a ground wire). AKC popularity: 74th of 130 breeds.

Physical Features

1. Males stand at least 32 inches and weigh at least 120 pounds. Females stand at least 30 inches and weigh at least 105 pounds.

2. His coat is coarse and rough with bushy eyebrows and beard, and needs to be brushed and combed twice a week. Straggly hairs should be scissored and dead hairs stripped twice a year. Usually he is fawn, red, gray, or brindle (brownish with black stripes and flecks). Solid white and solid black are less common.

3. His ears fold back against his head but prick up halfway when he is excited. His eyes are dark. His nose is black. His teeth meet in a scissors or level bite.

4. Faults: His nose should not be brown or flesh colored.

5. Disqualifications: He should not stand under the height specification.

Health Problems

He is a short-lived breed (eight to ten years) and susceptible to hip dysplasia, bloat, and bone tumors. Buy only from OFA-registered parents.

Cautions when Buying

Don't choose a timid Wolfhound puppy.

Norwegian Elkhound

For Experienced Owners Only
Good with Children if Raised with Children
Medium in Size
Thick Medium-Length Coat

	High	*Med*	*Low*
EXERCISE REQUIRED		✹	
TRIMMING/CLIPPING REQUIRED			✹
AMOUNT OF SHEDDING	✹		
ACTIVITY INDOORS		✹	
EASE OF TRAINING			✹
SOCIABILITY WITH STRANGERS		✹	

Temperament

This is an alert, bold, capable, confident dog whose independent character requires supervision and firm obedience training. He's a bundle of energy just waiting for the signal to go, yet when well trained he can control himself and be calm and dignified. He is so adaptable that he does fine in the city if exercised enough; still, he is boisterous, plays hard, and likes to be outdoors, especially in cold weather. Without time and effort, he can be very rambunctious; he must be taught not to pull on the leash or jump on people. Most are friendly with strangers, but some are reserved, and even the friendly ones can be protective, so he should be accustomed to people at an early age. He can be aggressive with other dogs. He may refuse commands and guard objects from family members who have not established leadership over him. Don't play aggressive games like wrestling or tug-of-war. He barks *a lot*, in a high-pitched, piercing voice.

History

This comrade to the Vikings once hunted bear and moose—not by attacking, but by cornering the prey and barking furiously until the huntsmen arrived. The breed is very old and developed with little influence from man. His name comes from the Norwegian *elg* (elk) *hund* (dog); Americans anglicized the *hund* into "hound." Today he is a companion but will probably still tree squirrels and bring bobcats to bay. AKC popularity: 44th of 130 breeds.

Physical Features

1. He stands 19½–20½ inches and weighs 48–55 pounds.
2. His coat is harsh with a dense undercoat, and should be brushed twice a week, but brush daily during shedding. He is shades of gray and silver, the hairs tipped with black. His muzzle, ears, and tail tip are always black. A puppy is born black and lightens as he matures.
3. His ears prick up. His eyes are dark brown. His nose is black. His teeth meet in a scissors bite. His tail curls tightly over his back.
4. Faults: He should not have yellow or brown shadings or white patches. He should not have light circles around his eyes. His tail should not be loosely curled.
5. Disqualifications: He should not be any color other than gray, such as black, white, red, or brown.

Health Problems

He is susceptible to hip dysplasia, eye problems, cysts, and kidney disease. Buy only from OFA-registered parents.

Cautions when Buying

BE CAREFUL. This is a strong-willed, independent breed, and if you buy from a poor breeder or raise the dog incorrectly, you could end up with a hardheaded Elkhound. Don't choose the boldest, noisiest, most independent Elkhound puppy.

Otterhound

For Experienced Owners Only
Good with Children
Large to Giant in Size
Long Coat

	High	Med	Low
EXERCISE REQUIRED		✦	
TRIMMING/CLIPPING REQUIRED			✦
AMOUNT OF SHEDDING		✦	
ACTIVITY INDOORS			✦
EASE OF TRAINING			✦
SOCIABILITY WITH STRANGERS		✦	

Temperament

He is utterly amiable and easygoing, but he needs so much free-space exercise, including swimming whenever possible, that he's completely unsuited to the city. He's energetic, and confinement will make him too rambunctious. He's usually fine with strangers, although some are standoffish. He can be a bit clumsy with small children. He can be aggressive with strange dogs and should be watched around small pets like cats and rabbits. Obedience training takes much time and effort, for he's very stubborn, but although he'll be slow to obey commands, he'll probably be good-natured about it. He must be taught not to pull on the leash, to stay put when told, and to come when called. He can be hard to housebreak. He has a loud, distinctive bay.

History

Kept in huge packs, he was once used to dispatch otters preying on the trout in popular English fishing streams. Today he is a companion but would probably still hunt opossum and raccoon. AKC popularity: 126th of 130 breeds.

Physical Features

1. He stands 22–27 inches and weighs 65–115 pounds.
2. His coat is 3–6 inches long, and shaggy with a woolly undercoat. His coat should be brushed every other day or he will be a matted mess. Colors include blue-and-white, sandy, gray, black-and-tan, and others.
3. His ears hang down; when pulled forward, they reach beyond his nose. His eyes and nose are black or brown. His teeth meet in a scissors bite. His feet are webbed to aid in swimming.
4. Faults: His coat should not be much longer than 6 inches.

Health Problems

He is susceptible to hip dysplasia, bloat, and hemophilia (bleeding). Buy only from OFA-registered or X-rayed parents who have been tested for VWD.

Cautions when Buying

Don't choose the most boisterous or independent Otterhound puppy.

Petit Basset Griffon Vendeen (Peh-TEET BAS-et Griff-ON Ven-DEE-en)

Fine for Novice Owners ✓
Good with Children ✓
Small in Size ✓
Wiry Coat ✓

	High	Med	Low
EXERCISE REQUIRED ✓		🦴	🦴
TRIMMING/CLIPPING REQUIRED ✓			🦴
AMOUNT OF SHEDDING ✓			🦴
ACTIVITY INDOORS	🦴		🦴
EASE OF TRAINING			🦴
SOCIABILITY WITH STRANGERS ✓	🦴		

Temperament

Despite his name, he is not Bassetlike but terrierlike: curious, busy, enthusiastic, reacting quickly to interesting scents and sudden movements, always looking for something to do. He's energetic but can adapt to the city if given long daily walks and frequent romps. He should not be let off the leash except in a safe, enclosed area, for he is a sniffer and an explorer, and he can take off. He is friendly with strangers and gets along well with other animals. He needs firm obedience training, for he is stubborn and independent—this is an engaging, hardy little dog, not a lapdog. He's an agile jumper, a digger, and a barker.

History

A photo in *Dog World* magazine in 1983 of a charming twelve-week-old puppy named "Alexander of Gebeba" was the first Petit Basset ever seen by most Americans, and his casual, rustic, shaggy appearance was so appealing that public inquiries poured in. He is a native of the French province of Vendee, a rugged region of dense brambly underbrush, where rabbit hunters needed sturdy little dogs with rough protective coats. The Grand (large) and Petit (small) varieties of Basset (low to the ground) Griffon (wiry-coated dog) fit the bill. Enthusiasts feel strongly that he should continue to be hunted in trials so that his working abilities do not become lost in favor of appearance. He is a companion and small-game hunter.

Physical Features

1. He stands 13–15 inches and weighs 35–45 pounds.
2. His coat is harsh and scruffy and should be brushed twice a week. He is mostly white, with lemon, orange, and/or grizzle markings.
3. His ears hang down. His eyes are brown. His nose is black.

Health Problems

He is susceptible to ear infections and possibly seizure problems in some lines.

Cautions when Buying

Don't choose the most independent Petit Basset puppy or a timid puppy.

Pharaoh Hound (FAIR-o)

Fine for Novice Owners
Good with Children
Medium to Large in Size
Smooth Coat

	High	Med	Low
EXERCISE REQUIRED	●		
TRIMMING/CLIPPING REQUIRED			●
AMOUNT OF SHEDDING		●	
ACTIVITY INDOORS			●
EASE OF TRAINING		●	
SOCIABILITY WITH STRANGERS		●	

Temperament

He is clean and well behaved, playful and athletic, light on his feet and a good jumper. He can be kept in the city if given regular running exercise, but he should not be let off the leash except in a safe, enclosed area—at full gallop he is extremely fast and can be a chaser. Most are cautious dogs, hesitantly investigating new people, new places, and new objects. To avoid real timidity, he should be accustomed to people and noises at an early age. He's fine with other dogs but should be watched around small pets like cats and rabbits. This sensitive dog responds well to obedience training that is persuasive and does not include a lot of jerking around. It's said that he blushes—that when he is excited or happy, his nose, ears, and eyes turn rosy pink.

History

This ancient breed was owned by Egyptian rulers and buried in his master's tomb. Called a sighthound because he uses his keen sight to locate prey, he once ran down gazelles and hares. Today he is a companion, sometimes participating in the sport of lure coursing (chasing a trash bag pulled along a ground wire). AKC popularity: 124th of 130 breeds.

Physical Features

1. He stands 21–25 inches and weighs 45–60 pounds.
2. His hard coat needs only a quick brushing once a week. He is tannish (most common) or reddish. Usually he has white markings on his chest and tail; sometimes he has white toes and a white snip on his face.
3. His ears prick up. His eyes are amber (brownish yellow) with a keen expression. His nose is flesh colored. His teeth meet in a scissors bite.
4. Faults: He should not have white flecking.
5. Disqualifications: He should not have a solid white spot on the back of his neck, shoulders, back, or sides.

Health Problems

He is sensitive to drugs (anesthetics, flea powders), so he should never be casually dosed with medication.

Cautions when Buying

Don't choose the most independent Pharaoh Hound puppy or a timid puppy.

Rhodesian Ridgeback

For Experienced Owners Only
Good with Older, Considerate Children
Large in Size
Smooth Coat

	High	Med	Low
EXERCISE REQUIRED	●		
TRIMMING/CLIPPING REQUIRED			●
AMOUNT OF SHEDDING		●	
ACTIVITY INDOORS			●
EASE OF TRAINING			●
SOCIABILITY WITH STRANGERS			●

Temperament

This most aggressive of the hounds might be okay in the city if given hard daily exercise, but he's much better in the suburbs or country where he has space. Reserved and protective with strangers, he should be accustomed to people at an early age. He can be aggressive with other dogs. This serious, strong-willed dog needs a firm, rigorous course of obedience training to keep him under control. He may refuse commands from family members who have not established leadership over him. Don't play aggressive games like wrestling or tug-of-war. Overall, he's a splendid, capable companion for experienced dog owners, but without ongoing exercise, supervision, and training, he can be hardheaded, aggressive, and rambunctious.

History

He comes from Rhodesia, Africa, and his name is derived from the peculiar ridge on his back: a raised strip of stiff hair that grows in the opposite direction of the rest of his coat. Originally a bird and game hunter, he was developed into a lion hunter, with the huntsmen following on horseback. Today he is a home guardian and companion. AKC popularity: 58th of 130 breeds.

Physical Features

1. He stands 24–27 inches and weighs 65–75 pounds.
2. He has a hard coat, which needs only a quick brushing once a week. He is light wheaten (golden tan) or dark wheaten (reddish tan). A little white on his chest and toes is okay.
3. His ears hang down. Light wheaten dogs have amber (brownish yellow) eyes and a brown nose; dark wheaten dogs have dark eyes and a black nose. His teeth meet in a level bite.
4. Faults: His nose should not be any color except black or brown.

Health Problems

He is susceptible to hip dysplasia and serious cyst problems. Buy only from OFA-registered parents.

Cautions when Buying

BE CAREFUL. This is a powerful, independent breed, and if you buy from a poor breeder or raise the dog incorrectly, you could end up with an overly dominant, aggressive Ridgeback. Don't choose the boldest, most energetic, or most independent Ridgeback puppy, and don't choose a timid puppy.

Feathered Saluki and Smooth Saluki (Sa-LOO-key)

These are varieties of the same breed.

For Experienced Owners Only
Good with Older, Considerate Children
Medium to Large in Size
Short Coat

	High	Med	Low
EXERCISE REQUIRED	●		
TRIMMING/CLIPPING REQUIRED			●
AMOUNT OF SHEDDING		●	
ACTIVITY INDOORS			●
EASE OF TRAINING			●
SOCIABILITY WITH STRANGERS			●

Temperament

He is a quiet, gentle aristocrat and not very demonstrative. Although calm in the house, he does best in the quiet suburbs or country where he can get running exercise and where he won't be startled by city sights and sounds. He's an agile jumper, and should not be let off the leash except in a safe, enclosed area—at full gallop he is extremely fast and can be a chaser. Reserved and often timid with strangers, he should be accustomed to people and noises at an early age. He's fine with other dogs but should be watched around small pets like cats and rabbits. Stubborn, skittish, and sensitive, he must be handled with much praise and little jerking around, but obedience training is wonderful for building his confidence and improving communication with him. He can bolt if startled or touched suddenly. He can be hard to housebreak. He should live with calm, quiet (but athletic) owners who will keep him exercised and relaxed.

History

An ancient breed from Egypt and Arabia, he is called a sighthound because he uses his keen sight to locate prey. He once ran gazelles down across open terrain, with trained falcons helping to distract the gazelle while the huntsmen followed on horseback. He was often buried in the tomb of his Egyptian master. The Mohammedans considered him the only sacred canine breed. Today he is a companion, sometimes participating in the sport of lure coursing (chasing a trash bag pulled along a ground wire). AKC popularity: 93rd of 130 breeds (but the Feathered variety is far more common).

Physical Features

1. Males stand 23–28 inches, females stand 20–25 inches. He weighs 40–60 pounds.
2. His coat is short and soft. The Feathered Saluki has longer silky hair on his legs, ears, and tail that should be combed once a week. Colors include white, cream, fawn, red, black-and-tan, and others.
3. His ears hang down. His eyes are dark or hazel with a dignified, gentle expression. His nose is black or brown. His teeth meet in a level bite.

Health Problems

He is sensitive to drugs (anesthetics, flea powders), so he should never be casually dosed with medication. He is also sensitive emotionally and may respond to stress with digestive upsets and skin allergies.

Cautions when Buying

Don't choose the most independent Saluki puppy, and don't choose a timid puppy.

Scottish Deerhound

Fine for Novice Owners
Good with Children
Giant in Size
Wiry Coat

	High	Med	Low
EXERCISE REQUIRED	✦		
TRIMMING/CLIPPING REQUIRED		✦	
AMOUNT OF SHEDDING			✦
ACTIVITY INDOORS			✦
EASE OF TRAINING		✦	
SOCIABILITY WITH STRANGERS		✦	

Temperament

He is calm, quiet, gentle, dignified, and mild mannered, but he needs so much running exercise that he belongs in the suburbs or country. However, he should not be let off the leash except in a safe, enclosed area—at full gallop he can be out of sight in seconds. He's sensible with strangers, although some are timid. He's fine with other dogs but should be watched around small pets like cats and rabbits. Mildly stubborn and slow to obey commands, he does eventually respond to obedience training that is as patient and gentle as he is. He's so easygoing that he's seldom a problem in any case. To come when called is the most important command this galloping, independent chaser should learn.

History

Called a sighthound because he uses his keen sight to locate prey, he once ran down 250-pound stags in the Scottish Highlands. In the Middle Ages he was owned by Highland chieftains. Today he is a companion, occasionally running down coyotes and jackrabbits, and participating in the sport of lure coursing (chasing a trash bag pulled along a ground wire). AKC popularity: 112th of 130 breeds.

Physical Features

1. Males stand at least 30 inches and weigh 85–110 pounds. Females stand at least 28 inches and weigh 75–95 pounds.
2. His coat is 3–4 inches long and ragged, with bushy eyebrows and a beard. He needs to be brushed and combed twice a week, and straggly hairs should be scissored and dead hairs stripped twice a year. He is dark blue-gray, light gray, sandy fawn, or brindle (brownish with black stripes and flecks). His ears and muzzle are black. Some white on his chest, toes, and tail is allowed.
3. His ears fold back against his head but prick up halfway when he is excited. His eyes are brown or hazel with a soft expression. His nose is black or blue. His teeth meet in a level bite.
4. Faults: His ears should not prick completely up.
5. Disqualifications: He should not have a white blaze on his head or a white neck collar.

Health Problems

He is a short-lived breed (ten years) and susceptible to bloat.

Cautions when Buying

Don't choose the most independent Deerhound puppy, and don't choose a timid puppy.

Whippet

Fine for Novice Owners
Good with Older, Considerate Children
Medium in Size
Smooth Coat

	High	Med	Low
EXERCISE REQUIRED		●	
TRIMMING/CLIPPING REQUIRED			●
AMOUNT OF SHEDDING		●	
ACTIVITY INDOORS			●
EASE OF TRAINING		●	
SOCIABILITY WITH STRANGERS			●

Temperament

He is clean, quiet, gentle, and playful. He's hardier than he looks, although he does love to curl up under blankets. He's fine in the city if taken for long daily walks and occasionally allowed to run. However, he should not be let off the leash except in a safe, enclosed area—he is the fastest dog of his weight, and he can be a chaser. Reserved and often timid with strangers, he should be accustomed to people and noises at an early age. He's fine with other dogs and cats, but should be watched around very small pets like rabbits and rodents. Obedience training is wonderful for building his confidence, but he's quite sensitive and only responds to calm, patient handling that includes much praise and little jerking around. He must learn to come when called. He can snap if startled or touched suddenly. Some Whippets develop nervous phobias, and some can be hard to housebreak.

History

Developed in England and bred down from the Greyhound and crossed with terriers, he was used as a "snapdog" in the cruel competition of snapping up rabbits in an enclosed arena. He is capable of speeds up to thirty-five miles per hour. It's a shame that Whippet racing is not as popular as Greyhound racing, for the former is far more humane because the racing Whippets double as pets and companions. AKC popularity: 61st of 130 breeds.

Physical Features

1. Males stand 19–22 inches, females stand 18–21 inches. He weighs 23–37 pounds.
2. His hard coat needs only a quick brushing once a week. Colors include white, tan, black, patched, brindle (brownish with black stripes and flecks), and others.
3. His ears fold back against his head but prick up halfway when he is excited. His eyes are dark with a keen expression. His nose is black. His teeth meet in a scissors bite.
4. Faults: His eyes should not be yellow.
5. Disqualifications: His eyes should not be blue. His teeth should not meet in an overshot bite of ¼ inch or more, nor any undershot bite. His coat should not be long. He should not stand more than ½ inch over or under the height specification.

Health Problems

He is a long-lived (fifteen years) breed. He is sensitive to drugs (anesthetics, flea powders), so he should never be casually dosed with medication. He is also sensitive emotionally and may respond to stress with digestive upsets. He should be protected from the cold.

Cautions when Buying

Don't choose a timid Whippet puppy.

The Working Group

Akita · Alaskan Malamute · Bernese Mountain Dog · Boxer
Bullmastiff · Doberman Pinscher · Giant Schnauzer · Great Dane
Great Pyrenees · Komondor · Kuvasz · Mastiff · Newfoundland
Portuguese Water Dog · Rottweiler
Longhaired Saint Bernard and Shorthaired Saint Bernard · Samoyed
Siberian Husky · Standard Schnauzer

Akita (a-KEE-ta)

For Experienced Owners Only
Good with Children if Raised with Children
Large to Giant in Size
Thick, Medium-Length Coat

	High	Med	Low
EXERCISE REQUIRED		●	
TRIMMING/CLIPPING REQUIRED			●
AMOUNT OF SHEDDING	●		
ACTIVITY INDOORS			●
EASE OF TRAINING	●		●
SOCIABILITY WITH STRANGERS			●

Temperament

He is calm, dignified, capable, generally easygoing. He is adaptable to the city if given enough daily exercise. Powerful, reserved, and protective, he should be accustomed to people at an early age. He can be aggressive with other dogs. He's assertive and strong willed, and needs a full course of obedience training from a firm, consistent hand, but he responds fairly well to it. Use much praise and not much repetition. He may refuse commands from family members who have not established leadership over him. Don't play aggressive games like wrestling or tug-of-war. Early socialization, proper exercise, and ongoing obedience training are necessities with an Akita. This breed must be treated with respect.

History

In the rugged mountains of Akita Prefecture, Japan, he was owned by the imperial aristocracy as a bear hunter and guard dog. At a Tokyo rail station there is a statue honoring the faithful Akita "Hachiko," who waited nine years at the station for his master who had died at work. He is a home guardian and companion. AKC popularity: 37th of 130 breeds.

Physical Features

1. Males stand 26–28 inches, females stand 24–26 inches. He weighs 75–110 pounds.
2. His coat is 1–2 inches long and harsh, with a dense undercoat, and needs brushing once a week. Colors include fawn, brindle, pinto (white with balanced colored patches covering two-thirds of his body), solid white, and others. Often there is a mask or blaze on his face.
3. His ears prick up and incline forward. His eyes are dark brown and triangular. His nose is black, but may be brown on white dogs. His teeth meet in a scissors or level bite. His tail is carried over his back in a three-quarter, full, or double curl.
4. Disqualifications: His nose should not be two-toned. His ears should not hang down or tip over. His tail should not be uncurled when he is in motion. His teeth should not meet in a noticeably overshot or undershot bite. He should not stand more than 1 inch under the height specification.

Health Problems

He is susceptible to hip dysplasia, patella problems (bad knees), popping hocks, thyroid problems, and PRA. Buy only from OFA- and CERF-registered parents.

Cautions when Buying

BE CAREFUL. This is a powerful, independent breed, and if you buy from a poor breeder or raise the dog incorrectly, you could end up with an overly dominant, aggressive Akita. Don't choose the boldest or most independent Akita puppy, and don't choose a timid puppy.

Alaskan Malamute

For Experienced Owners Only
Good with Older, Considerate Children
Large in Size
Thick, Medium-Length Coat

	High	Med	Low
EXERCISE REQUIRED		✦	
TRIMMING/CLIPPING REQUIRED			✦
AMOUNT OF SHEDDING	✦		
ACTIVITY INDOORS		✦	
EASE OF TRAINING			✦
SOCIABILITY WITH STRANGERS	✦		

Temperament

Some remain playful puppies all their lives, others mature into calm, dignified adults. He likes the cold outdoors and belongs in the suburbs or country. He can be boisterous and needs a good amount of daily exercise. Too much confinement or isolation makes him restless and destructive—some females will dig den holes in the sofa. Most are very friendly with strangers, but a few are reserved, and this is a big breed, so he should be accustomed to people at an early age. He can be aggressive with other dogs; he should be watched around small pets like cats and rabbits, and must be supervised outdoors, for he is an explorer who could run deer and molest livestock. Powerful and independent, he needs much firm obedience training (especially the commands *heel*, *down*, and *come*). Male Malamutes can be very dominant, and unless neutered and well-trained, may refuse commands from family members who have not established leadership over them. He howls a lot and can relandscape your yard with his digging. When well-trained, exercised, and supervised, this breed makes an impressive companion.

History

His name comes from the Mahlemut tribe, who used him to pull sledges. Today he is a pack dog, freight dog, and companion. He's not built for speed, so he's not a racer, but champion weight-pullers pull over 1000 pounds. AKC popularity: 40th of 130 breeds.

Physical Features

1. He stands 23–25 inches and weighs 75–110 pounds.
2. His coat is 1–2 inches long and coarse, with a dense undercoat, and should be brushed twice a week—daily when he's shedding. He should be black, Alaskan seal (black with a creamy undercoat), sable (black with a red undercoat), wolf gray (gray with a gray/creamy undercoat), silver (light gray with a white undercoat), or red; he has white markings on his face, neck, chest, and legs. Rarely, he is solid white.
3. His ears prick up. His eyes are brown, but may be lighter in red dogs. His nose is black, but may be flesh colored on red dogs. His teeth meet in a scissors bite. His tail is carried over his back like a waving plume.
4. Disqualifications: His eyes should not be blue. (Identification note: the Siberian Husky can have blue eyes.)

Health Problems

He is susceptible to hip dysplasia, zinc-deficiency skin disorders, thyroid problems, and achondrodysplasia (akon-dro-dis-PLAY-zhia), a congenital dwarfism where a puppy is born with a normal head and body but short, warped legs. With his thick coat, he doesn't like the heat. Buy only from OFA-registered and ChD- (achondrodysplasia) cleared parents.

Cautions when Buying

BE CAREFUL. This is a powerful, independent breed, and if you buy from a poor breeder or raise the dog incorrectly, you could end up with a dominant boss-dog Malamute. Don't choose the boldest, most energetic, or most independent Malamute puppy.

Bernese Mountain Dog

Fine for Novice Owners
Good with Children
Large to Giant in Size
Thick, Medium-Length Coat

	High	Med	Low
EXERCISE REQUIRED		●	
TRIMMING/CLIPPING REQUIRED			●
AMOUNT OF SHEDDING		●	
ACTIVITY INDOORS			●
EASE OF TRAINING	●	●	
SOCIABILITY WITH STRANGERS		●	

Temperament

He is steady tempered, gentle, and easygoing, but because he is a big, heavy dog who needs daily exercise and loves the outdoors, he belongs in the suburbs or country. Some are friendly with strangers, some are reserved. There is a slight tendency toward timidity, so he should be accustomed to people and noises at an early age. He gets along well with other animals. Responsive to obedience training in a patient, good-natured way, this sensitive, willing breed should be handled kindly, with much praise and encouragement; some Bernese, especially the less-bulky females, can excel in obedience. He enjoys pulling carts and sleds. The only real behavioral problem, especially in female Bernese, may be a tendency to attach too strongly to one person.

History

This is an old breed; his ancestors were brought to Switzerland by invading Roman legions. His name comes from the Swiss canton of Berne, where he was a farm guardian and cart puller. There are several varieties of Swiss Mountain Dog in Europe, but the Bernese is the only one fully recognized by the AKC; the shorter-coated Greater Swiss Mountain Dog is partially recognized. Today the Bernese is a companion. AKC popularity: 76th of 130 breeds.

Physical Features

1. Males stand 24½–27½ inches and weigh 75–110 pounds. Females stand 23–25½ inches and weigh 65–90 pounds.
2. His coat is slightly wavy and silky and should be brushed twice a week. He is jet black with white markings on his face, muzzle, chest, feet, and tail tip, and with rust markings over each eye, on his cheeks, on each side of his chest, on each leg, and under his tail.
3. His ears hang down. His eyes are dark brown with a gentle expression. His nose is black. His teeth meet in a scissors bite.
4. Faults: His legs should not be white. He should not have white markings around his neck that form a collar.
5. Disqualifications: His eyes should not be blue. He should not be any color other than black/rust/white.

Health Problems

He is short-lived (ten years) and susceptible to hip and elbow dysplasia, bloat, cancer, and some eyelid abnormalities. With his thick black coat, he doesn't like the heat. Buy only from parents whose hips and elbows are OFA-registered.

Cautions when Buying

Note: some lines of Bernese are huge and some are more medium in size, so ask the breeder about his sizes. Don't choose a timid Bernese puppy.

Boxer

Fine for Novice Owners
Good with Children
Large in Size
Smooth Coat

	High	Med	Low
EXERCISE REQUIRED		●	
TRIMMING/CLIPPING REQUIRED			●
AMOUNT OF SHEDDING		●	
ACTIVITY INDOORS		●	
EASE OF TRAINING			●
SOCIABILITY WITH STRANGERS		●	

Temperament

This good-natured dog is usually bouncy and playful, although some adults are more deliberate and dignified. He has been called an "honest" dog because his face expresses his emotions. He can adapt to the city if given enough daily exercise to work off his exuberance. He likes to participate in games and activities. He may be friendly, reserved, protective, or timid with strangers, so he should be accustomed to people at an early age. He can be aggressive with strange dogs. He needs calm, firm obedience training because he's strong and stubborn, but he is sensitive to correction, so don't jerk him endlessly. *Heel, down,* and *come* are important commands for a Boxer. He may refuse commands from family members who have not established leadership over him. Don't play aggressive games like wrestling or tug-of-war. He wheezes.

History

Descended from sixteenth-century European bulldogs, he was developed in Germany as a dogfighter, bullbaiter, and police dog. His name comes from his habit of striking out with his front feet when playing or fighting. Today the ferocity has been bred out of him and he is a home guardian and companion. AKC popularity: 21st of 130 breeds.

Physical Features

1. He stands 21–25 inches and weighs 60–75 pounds.
2. His hard coat needs only a quick brushing once a week. He is fawn (light tan), mahogany brown, or brindle (brownish with black stripes). White markings on his face, chest, neck, and feet are desirable. His muzzle should be black, but white may replace some of the black.
3. His ears must be cropped for show, but may be left hanging in pets. His eyes are dark brown. His nose is black. He is one of the few breeds whose teeth meet in an undershot bite. His tail is docked.
4. Faults: His teeth or tongue shouldn't show when his mouth is closed.
5. Disqualifications: He should not be any color other than fawn, mahogany, or brindle, such as white or black. He should not have white markings on more than one-third of his body.

Health Problems

He is a short-lived breed (ten years) and susceptible to tumors, digestive problems, heart disease, corneal ulcers, and bloat. With his pushed-in face, he is very sensitive to hot, stuffy conditions and heatstroke—don't leave him in a closed car.

Cautions when Buying

THERE ARE MANY BOXERS AROUND—BE VERY CAREFUL. If you buy this breed from a poor breeder, you could wind up with a sickly, nervous, aggressive, or hyperactive Boxer. Don't choose the boldest, most energetic, most independent Boxer puppy, and don't choose a timid puppy.

Bullmastiff

For Experienced Owners Only
Good with Children if Raised with Children
Giant in Size
Smooth Coat

	High	Med	Low
EXERCISE REQUIRED		●	
TRIMMING/CLIPPING REQUIRED			●
AMOUNT OF SHEDDING		●	
ACTIVITY INDOORS			●
EASE OF TRAINING			●
SOCIABILITY WITH STRANGERS		●	

Temperament

He is almost always docile and mild mannered, but he is also serious, and once aroused will seldom back down. He belongs in the suburbs or country, but he could adapt to the city if he has some space and exercise. He is sensible with strangers, but this powerful, protective breed should be accustomed to people at an early age. He can be aggressive with other dogs. Although stubborn and tremendously strong, he will respond to early, firm, consistent obedience training that includes cheerful praise and encouragement. He must be taught not to pull on the leash. He may refuse commands from family members who have not established leadership over him. Don't play aggressive games like wrestling or tug-of-war, and anyone who hits a Bullmastiff is asking for trouble. He drools. Exercise, early socialization, and full obedience training are necessities with this breed.

History

He was developed in England as a cross of 60% Mastiff and 40% Bulldog. A night watchdog on large estates and game preserves, he was called "the gamekeeper's nightdog." He attacked poachers and pinned them to the ground but didn't bite them. At that time, dark brindle was the preferred color, because dark dogs were camouflaged at night; today fawn is more popular. He is a home guardian and companion. AKC popularity: 69th of 130 breeds.

Physical Features

1. He stands 24–27 inches and weighs 100–130 pounds.
2. His hard coat needs only a quick brushing once a week. He is fawn, red, or brindle. His muzzle is always black. A small white spot on his chest is allowed.
3. His ears hang down. His eyes are dark. His nose is black. His teeth meet in a level or undershot bite.
4. Faults: He should not have any white markings, except on his chest. He should not have a very short muzzle like that of a Bulldog—the length of his muzzle should take up a full one-third of his entire head length.

Health Problems

He is a short-lived breed (ten years) and susceptible to hip dysplasia, bloat, serious tumors, and eyelid abnormalities. Buy only from OFA-registered parents.

Cautions when Buying

BE CAREFUL. This is an extremely powerful breed, and if you buy from a poor breeder or raise the dog incorrectly, you could end up with an overly dominant, aggressive Bullmastiff. Don't choose the boldest, most energetic, or most independent Bullmastiff puppy, and don't choose a timid puppy.

Doberman Pinscher (PIN-sher)

For Experienced Owners Only
Good with Children if Raised with Children
Large in Size
Smooth Coat

	High	Med	Low
EXERCISE REQUIRED		●	
TRIMMING/CLIPPING REQUIRED			●
AMOUNT OF SHEDDING		●	
ACTIVITY INDOORS	●	●	
EASE OF TRAINING	●	●	
SOCIABILITY WITH STRANGERS			●

Temperament

Some are bold and outgoing, some are sweet and mellow, some are nervous or suspicious. Many are loving family dogs, but some are one-person dogs. He can adapt to the city if given daily physical and also mental exercise (thorough obedience training), but he is restless without something to do. Reserved and protective, he must be accustomed to people at an early age. He can be aggressive with other dogs. This capable dog works beautifully in obedience, but he is very sensitive to correction, responding only to calm training and absolutely no abuse. He may bite if hit or startled. He may refuse commands from family members who have not established leadership over him. Don't play aggressive games like wrestling or tug-of-war. Ongoing companionship, socialization, and activity are essential in keeping this breed relaxed and controlled.

History

He was developed in Germany by Louis Dobermann from various guard dogs and terriers. *Pinscher* means "terrier," but today's Doberman bears no physical or temperamental resemblance to a terrier, so in Germany, the "Pinscher" has been dropped. He is a guard dog, military dog, Schutzhund dog, home guardian, and companion. AKC popularity: 19th of 130 breeds.

Physical Features

1. Males stand 26–28 inches, females stand 24–26 inches. He weighs 60–85 pounds.

2. His hard coat needs only a quick brushing once a week. Usually he's black or red (dark brown) with rust markings above his eyes, on his muzzle, throat, chest, legs and feet, and below his tail. He can also be fawn (called Isabella) or steel blue, with the same rust markings. A white spot on his chest is allowed.

3. His ears may be cropped or left hanging. His eyes are dark in black dogs, lighter in others. His nose is black on black dogs, brown on red dogs, gray on blue dogs, tan on fawn dogs. His teeth meet in a scissors bite. His tail is docked to the second joint.

4. Disqualifications: His teeth should not meet in a noticeably overshot or undershot bite. He should not have more than three missing teeth. He should not be any color other than the four allowed, such as white.

Health Problems

He is susceptible to hip dysplasia, von Willebrand's Disease, bloat, immune-deficiency disorders, severe heart disease, and thyroid and liver disorders. Buy only from OFA-registered, VWD-tested, cardio-normal (heart-tested) parents.

Cautions when Buying

THERE ARE MANY DOBERMANS AROUND—BE VERY CAREFUL. Show Dobes are often slender and elegant, while working Dobes are often from aggressive German lines. Look for a breeder who strives for good temperament, good health, good conformation, and some working ability in obedience competition. If you buy from a poor breeder or raise the dog incorrectly, you could wind up with a sickly or aggressive Doberman. Don't choose the boldest or most independent Doberman puppy or a timid puppy.

Giant Schnauzer (SCHNOW-zer)

For Experienced Owners Only
Good with Older, Considerate Children
Large in Size
Wiry Coat

	High	Med	Low
EXERCISE REQUIRED	●		
TRIMMING/CLIPPING REQUIRED	●		
AMOUNT OF SHEDDING			●
ACTIVITY INDOORS		●	
EASE OF TRAINING			●
SOCIABILITY WITH STRANGERS			●

Temperament

He is keen, bold, reliable, and responsible. This energetic breed does best in the suburbs or country with an athletic owner who will take him out to run. Many are too spirited for small children. Reserved and protective, he should be accustomed to people at an early age. He can be aggressive with other dogs. He's independent, capable, and trusts his own judgment, but he responds well to experienced obedience training that is fair, consistent, and not harsh. Never hit a Schnauzer or he could retaliate. He may refuse commands from family members who have not established leadership over him. Don't play aggressive games like wrestling or tug-of-war. Some Giants are much more tranquil than others, and some are a bit timid or skittish.

History

He drove cattle in the German region of Bavaria, where all three Schnauzers (Miniature, Standard, and Giant) originated. They are separate breeds, the Standard being the oldest, the Miniature the newest, and the Giant resulting from crosses of herding dogs with (probably) the black Great Dane and (perhaps) the Bouvier des Flandres. Today he is a police dog, Schutzhund dog, home guardian, and companion. AKC popularity: 67th of 130 breeds.

Physical Features

1. Males stand 25½–27½ inches, females stand 23½–25½ inches. He weighs 70–95 pounds.
2. His coat is harsh, with bushy eyebrows and a beard, and needs brushing and combing twice a week. He needs scissoring and his coat needs shaping every three months. For pets, coat shaping means electric clipping; for show dogs, it means stripping (plucking) the dead hairs out one by one, because electric clipping softens the coat and fades the color. He is usually solid black (a white spot on his chest is allowed), but he also may be pepper-and-salt (gray with a liberal sprinkling of black and white hairs).
3. His ears may be cropped or left folded forward naturally. His eyes are dark brown with a keen expression. His nose is black. His teeth meet in a scissors bite. His tail is docked to the second or third joint.
4. Disqualifications: His teeth should not meet in an overshot or undershot bite. He should have no color markings other than those listed.

Health Problems

He is susceptible to hip dysplasia. Buy only from OFA-registered parents.

Cautions when Buying

BE CAREFUL. This is a strong, independent breed, and if you buy from a poor breeder or raise the dog incorrectly you could end up with an overly dominant, aggressive Giant Schnauzer. Don't choose the boldest, most energetic, or most independent Giant puppy, and don't choose a timid puppy.

Great Dane

Fine for Novice Owners
Good with Children
Giant in Size
Smooth Coat

	High	Med	Low
EXERCISE REQUIRED		●	
TRIMMING/CLIPPING REQUIRED			●
AMOUNT OF SHEDDING		●	
ACTIVITY INDOORS			●
EASE OF TRAINING		●	
SOCIABILITY WITH STRANGERS		●	

Temperament

He is usually gentle, easygoing, and mild mannered, but can be spirited. He does best in the suburbs or country—his big body must be given daily exercise to stay fit, whether he appears to want it or not. Some are friendly with strangers, some are reserved, some are timid, so he must be accustomed to people and noises at an early age. He likes children but may be too clumsy for toddlers. He's usually good with other animals, but some can be aggressive with other dogs. This huge breed must have a full course of obedience training to be well mannered, but he's a willing dog who should be handled with calm praise and encouragement. Harsh discipline and leash jerking make him skittish. He must be taught not to pull on the leash or jump on people. A dominant Dane could refuse commands from family members who have not established leadership over him. Don't play aggressive games like wrestling or tug-of-war. Above all, this breed needs space and companionship.

History

Despite his name, he originated in Germany—in Europe he is correctly called German Mastiff. He was used for guarding, and for hunting the savage European wild boar. His nickname is "The Apollo of Dogs." Today he is a companion. AKC popularity: 36th of 130 breeds.

Physical Features

1. Males stand at least 30 inches, females stand at least 28 inches. He weighs 120–150 pounds.

2. His hard coat needs only a quick brushing once a week. He is solid fawn (tan) with a black muzzle, solid black, solid steel blue, brindle (brownish with black stripes and flecks), or harlequin (white with irregular black patches).

3. His ears may be cropped or left hanging. His eyes are dark but may be lighter in black or blue Danes, and may be blue or two different colors in harlequin Danes. His nose is black but may be black-spotted on harlequin Danes. His teeth meet in a scissors bite.

4. Disqualifications: He should not stand under the height specification. He should not be any color other than the five listed, such as solid white, gray, spotted, or blue merle. He should not have markings such as a white belly, white collar, white feet, or white legs—these Danes are often advertised as "rare Bostons," and are recognized in Canada but not in the United States.

Health Problems

He is a short-lived breed (ten years) and susceptible to hip dysplasia, bloat, bone cancer, heart disease, and tumors. Buy only from OFA-registered parents.

Cautions when Buying

BE CAREFUL. If you buy this breed from a poor breeder or raise the dog incorrectly, you could end up with a sickly, hyperactive, nervous, or aggressive Dane. Don't choose the most independent Dane puppy, and don't choose a timid puppy.

Great Pyrenees

For Experienced Owners Only
Good with Children
Giant in Size
Thick Medium-Length Coat

	High	Med	Low
EXERCISE REQUIRED	●		
TRIMMING/CLIPPING REQUIRED			●
AMOUNT OF SHEDDING	●		
ACTIVITY INDOORS			●
EASE OF TRAINING			●
SOCIABILITY WITH STRANGERS			●

Temperament

He's a calm, serious, purposeful dog, quiet indoors but in need of so much space and exercise that he belongs in the suburbs or country. He loves the cold outdoors. This is a reserved, protective breed who occasionally becomes strongly attached to one person, so he should be accustomed to people at an early age. Males can be aggressive with other male dogs. Obedience training takes time and effort, for he is very independent (especially the male) and trusts his own judgment. He may refuse commands from family members who have not established leadership over him. Don't play aggressive games like wrestling or tug-of-war. Early socialization, ongoing companionship, and proper training and supervision are necessities with a Great Pyrenees.

History

He is an old breed from the high, isolated Pyrenean mountains between France and Spain. He guarded sheep flocks against prowling wolves and bears, and in the French medieval period was a royal court dog and guardian of the nobility. In World War I he was a pack dog and runner of contraband over the Franco-Spanish border. Today in his homeland he still guards livestock and mountain cottages, but in America he is a companion, a cart dog, and occasionally a pack dog on ski trips. AKC popularity: 51st of 130 breeds.

Physical Features

1. Males stand 27–32 inches, females stand 25–29 inches. He weighs 90–125 pounds.
2. His coat is coarse with a dense undercoat and needs to be brushed twice a week, but brush daily when he's shedding. He is solid white or mostly white with a few gray or tan markings.
3. His ears hang down. His eyes are dark brown with a kindly expression. His nose is black. He must have one dewclaw on each front leg and two dewclaws on each hind leg.

Health Problems

He is a short-lived breed (ten years) and susceptible to hip dysplasia, eyelid abnormalities, and bloat. Buy only from OFA-registered parents.

Cautions when Buying

BE CAREFUL. This is a powerful, independent breed, and if you buy from a poor breeder or raise the dog incorrectly, you could end up with an overly dominant, aggressive Great Pyrenees. Don't choose the boldest or most independent Pyr puppy or a timid puppy.

Komondor (KOM-on-door)

For Experienced Owners Only
Good with Older, Considerate Children
Giant in Size
Long Coat

	High	Med	Low
EXERCISE REQUIRED	✦		
TRIMMING/CLIPPING REQUIRED			✦
AMOUNT OF SHEDDING			✦
ACTIVITY INDOORS	✦	✦	
EASE OF TRAINING			✦
SOCIABILITY WITH STRANGERS			✦

Temperament

Although playful as a puppy, he matures into a calm, serious, self-reliant adult (at two or three years) who needs much supervision and control. This robust dog belongs in the suburbs or country, for he needs space and exercise and likes the cold outdoors. Reserved and extremely protective, he should be thoroughly accustomed to people at an early age. He can be aggressive with other dogs. This responsible, capable breed trusts his own judgment and makes his own decisions, but he will respect a strong, firm, experienced owner who gives him a full, rigorous course of obedience training. He may refuse commands from family members who have not established leadership over him. Don't play aggressive games like wrestling or tug-of-war. He can bite if teased or provoked. Don't let him off the leash except in a safe, enclosed area. If he lives in a busy neighborhood he can be noisy as he announces every strange person or sound. Ongoing socialization, exercise, supervision, and training are essential with a Komondor.

History

This imposing dog, described as a huge walking string mop, was developed on the Hungarian plains as a rugged guardian of sheep flocks; the actual herding was left to smaller dogs. Today in Hungary he is still a guardian who lives out in the open; in America, he is a home guardian and companion. AKC popularity: 105th of 130 breeds.

Physical Features

1. He stands 25–30 inches and weighs 85–125 pounds.
2. The wiry hairs of his outercoat fuse with the wooly hairs of his undercoat to form real felt cords, which must occasionally be separated; when the dog is shampooed, the cords must be rinsed well, and drying takes a long time. In pet dogs, you may want to trim the coat short for neatness, once or twice a year. A puppy's coat falls into the proper cords by two years of age. He is solid white, preferably with gray skin.
3. His ears hang down. His eyes are dark brown. His nose is black, dark brown, or dark gray. His teeth meet in a scissors or level bite.
4. Faults: His ears should not prick up. He should not have pink pigment on his nose, lips, pads of his feet, or around his eyes. His eyes should not be light-colored or yellowish.
5. Disqualifications: His eyes should not be blue-white. His tail should not be a bobtail. His nose should not be flesh-colored. The hair on his head and legs should not be short. He should not be any color other than white. If he is two years old or more, he cannot go into the show ring with an uncorded coat.

Health Problems

He is susceptible to hip dysplasia, bloat, and skin allergies. Buy only from OFA-registered parents.

Cautions when Buying

BE CAREFUL. This is a powerful, independent breed, and if you buy from a poor breeder or raise the dog incorrectly, you could end up with an aggressive Komondor. Don't choose the boldest, most energetic, or most independent Komondor puppy, and don't choose a timid puppy.

Kuvasz (KOO-vahz)

For Experienced Owners Only
Good with Older, Considerate Children
Giant in Size
Thick Medium-Length Coat

	High	Med	Low
EXERCISE REQUIRED	🦴		
TRIMMING/CLIPPING REQUIRED			🦴
AMOUNT OF SHEDDING	🦴		
ACTIVITY INDOORS		🦴	
EASE OF TRAINING			🦴
SOCIABILITY WITH STRANGERS			🦴

Temperament

He is a bold, determined, spirited dog who needs much supervision and control. He needs space and exercise and likes the cold outdoors, so he belongs in the suburbs or country. This is not a backyard breed—he must have plenty of attention and companionship. He is easily bored, and then he can be destructive. This territorial breed must have a fenced yard. A reserved and protective breed, he should be thoroughly accustomed to people at an early age. He can be aggressive with other dogs. He needs a full, rigorous course of obedience training and a strong, firm owner, for he is independent and headstrong. He may refuse commands from family members who have not established leadership over him. Don't play aggressive games like wrestling or tug-of-war. Ongoing socialization, exercise, supervision, and training are necessities with this breed. *Note:* Some Kuvasz lines are much more mellow than others.

History

His ancestors came from Tibet, but he was developed in Hungary as a guardian of the aristocracy. Rulers kept huge, commanding Kuvaszok (the plural of Kuvasz) as personal bodyguards. His name comes from the Turkish word *kawasz*, which means "armed guard of the nobility." Eventually the breed was used by the village commoners as a livestock guardian. Today he is a home guardian and companion. AKC popularity: 90th of 130 breeds.

Physical Features

1. Males stand 28–30 inches and weigh 100–120 pounds. Females stand 26–28 inches and weigh 70–100 pounds.
2. His coat is straight or very wavy with a dense undercoat and with very short hair on his head, ears, and paws. His coat needs brushing twice a week, but brush daily when he's shedding. He is solid white, his skin heavily pigmented with gray or black.
3. His ears hang down. His eyes are dark brown. His nose is black. His teeth meet in a scissors or level bite.
4. Disqualifications: His teeth should not meet in an overshot or undershot bite. He should not stand more than 2 inches under the height specification. He should not be any color other than white.

Health Problems

He is susceptible to hip dysplasia, bloat, skin problems, and allergies. Buy only from OFA-registered parents. He does not do well in high humidity or high temperatures.

Cautions when Buying

BE CAREFUL. This is a powerful, independent breed, and if you buy from a poor breeder or raise the dog incorrectly, you could end up with an overly dominant, aggressive Kuvasz. Don't choose the boldest, most energetic, or most independent Kuvasz puppy, and don't choose a timid puppy.

Mastiff

For Experienced Owners Only
Good with Children
Giant in Size
Smooth Coat

	High	Med	Low
EXERCISE REQUIRED		●	
TRIMMING/CLIPPING REQUIRED			●
AMOUNT OF SHEDDING		●	
ACTIVITY INDOORS			●
EASE OF TRAINING		●	
SOCIABILITY WITH STRANGERS		●	

Temperament

He is calm, easygoing, and dignified. He belongs in the suburbs or country because his massive body needs stretching room and regular exercise whether he seems to want it or not. He also needs companionship—he's not a dog to be relegated to the backyard. Because of his size and ancestry, he does need supervision and control, but he is harder to provoke or arouse than is the Bullmastiff. Most are friendly with strangers, some are reserved, and he should learn his proper place in society and be accustomed to people at an early age. Usually he's fine with other animals, but some can be aggressive with other dogs. He likes children but may be too ponderous for toddlers. Although mildly stubborn, he is a good-natured dog who responds well to patient obedience training that consists of much praise and little correction. Anyone who hits a Mastiff is asking for trouble. He drools and wheezes. Early socialization and training, and ongoing companionship, exercise, and control are necessities with a Mastiff.

History

He was a war dog with Caesar's Roman legions and a gladiator in the Roman arena, fighting dogs, bulls, and lions. In England he was a village guardian loosed at night to control prowling wolves. His name should really be Old English Mastiff, for the word "mastiff" means a *type* of dog—a giant fighting dog with a heavy head—rather than a specific breed. Today he is a home guardian and companion. AKC popularity: 54th of 130 breeds.

Physical Features

1. Males stand at least 30 inches, females stand at least 27½ inches. He weighs 170–200 pounds.
2. His hard coat needs only a quick brushing once a week. He is fawn, apricot, or brindle (brownish with black stripes and flecks). His muzzle and ears are black.
3. His ears hang down. His eyes are brown with a kindly expression. His nose is dark. His teeth meet in a scissors or undershot bite.

Health Problems

He is a short-lived breed (ten years) and susceptible to hip dysplasia and eyelid abnormalities. Buy only from OFA-registered parents.

Cautions when Buying

BE CAREFUL. This is a giant breed and if you buy from a poor breeder or raise the dog incorrectly, you could end up with a dominant, aggressive Mastiff. Don't choose the boldest or most independent Mastiff puppy, and don't choose a timid puppy.

Traits I'm look f m a dog

1 Exp. req. — Fine for novice
2. W children — Good w children
3. Size — Littles/Small
4. Coat — smooth/short/wiry-curly
5. Exerci req. — Low/med.
6. Trim/clip req. — Low/med./High
7. Shed — Low
8. Act. In. — Low/med.

Peh/High/med

BENJAMIN FRANKLIN SCHOOL
Westfield, NJ   07090

January 26, 1995

FROM THE PRINCIPAL.:

Since the bond referendum defeat in December, there have been many rumors and a good deal of misinformation regarding plans for the 1995-96 school year here at Franklin. It is very difficult to give absolutes this early in the planning stage. However, after reviewing current class enrollments and room availability, it appears that the following changes will be proposed for next year to allow us to cope with our increasing student enrollments.

1. Our current five sections of first grade with a total enrollment of 104 children will be reduced to four sections of second grade. Historically, we have increased enrollment slightly from first to second grade so I anticipate this will create classes that average between 26 and 28 children.

2. Our current third grade consisting of four sections with a total of 81 students, will be consolidated to three sections for fourth grade with class sizes averaging between 26 and 28 children.

Newfoundland (NEWF-und-lund)

Fine for Novice Owners
Good with Children
Giant in Size
Thick Medium-Length Coat

	High	Med	Low
EXERCISE REQUIRED		🦴	
TRIMMING/CLIPPING REQUIRED			🦴
AMOUNT OF SHEDDING	🦴		
ACTIVITY INDOORS			🦴
EASE OF TRAINING		🦴	
SOCIABILITY WITH STRANGERS	🦴		

Temperament

He is calm, quiet, patient, dignified, and reliable, but he belongs in the suburbs or country (preferably in a cold climate) where there's space for him. He needs long daily walks, occasional runs, and swimming whenever possible. He likes to be outdoors, but he's not a backyard breed—he's sociable and needs companionship. He is sensibly friendly with strangers, and although not a natural guardian of property, some Newfs will protect their people. Some males are aggressive with strange male dogs. He's mildly independent but very responsive to patient obedience training (females are especially willing to please). This sweet-tempered dog should be taught with much praise and encouragement—harsh jerking makes him shy. He's a slow-moving dog—give him time to respond and don't expect the quick precision of a Toy Poodle. He pants a lot, drinks a lot, and drools a lot.

History

In Newfoundland, Canada, he hauled fishermen's nets, carried lifelines to shipwrecked vessels, pulled carts, and carried packs. Lord Byron wrote an epitaph to the quiet heroics of "Boatswain" (BO-sun): "Near this spot are deposited the remains of one who possessed beauty without vanity, strength without insolence, courage without ferocity, and all the virtues of man without his vices." Today he is still a water rescue dog and companion. AKC popularity: 53rd of 130 breeds.

Physical Features

1. Males stand about 28 inches and weigh 130–150 pounds. Females stand about 26 inches and weigh 110–130 pounds.
2. His coat is coarse with a dense undercoat. He should be brushed twice a week, but brush daily when he's shedding. He's usually solid black (perhaps with a little white or brown on his chin, chest, toes, or tail tip), or white with black patches (called Landseer). Solid bronze (deep reddish brown) and solid gray are less common.
3. His ears hang down. His eyes are dark brown with a soft expression. His nose is usually black, but on bronze dogs, it's brown. His teeth meet in a scissors or level bite. His feet are webbed to aid in swimming.
4. Disqualifications: He should not be any color other than those listed above.

Health Problems

He is a short-lived breed (ten years) and susceptible to hip dysplasia, eyelid abnormalities, and heart disease. Buy only from OFA-registered parents. With his thick black coat he doesn't like the heat.

Cautions when Buying

Don't choose a timid Newf puppy.

Portuguese Water Dog

For Experienced Owners Only
Good with Children
Medium in Size
Curly Coat (But see Physical Features 2.)

	High	Med	Low
EXERCISE REQUIRED		•	
TRIMMING/CLIPPING REQUIRED	•		•
AMOUNT OF SHEDDING			•
ACTIVITY INDOORS		•	
EASE OF TRAINING		•	
SOCIABILITY WITH STRANGERS		•	

Temperament

Emotionally he's calm and sensible; physically he's lively and spirited, ready for any activity. He's so adaptable that he does fine in the city if given physical exercise (long walks, occasional runs, and swimming whenever possible) and also mental exercise (obedience training). He's usually friendly with strangers, but some are a bit timid, so he should be accustomed to people and noises at an early age. He's good with other animals if raised around other animals. Although independent and a bit sensitive, he's nicely responsive to firm yet gentle obedience training. PWD puppies and adolescents are notorious chewers. You must stay on top of this smart, capable breed.

History

He once worked along the rugged coast of Portugal herding fish into nets, retrieving equipment that fell overboard, diving after escaping fish, carrying messages from ship to ship, and riding codfish trawlers out to sea. Today in the United States he is a companion. AKC popularity: 92nd of 130 breeds.

Physical Features

1. Males stand 20–23 inches, females about 17–21 inches. He weighs 35–60 pounds.
2. His coat is either medium in length and wavy or short and curly and needs to be brushed twice a week. His coat may be clipped into a lion style (face and hindquarters completely shaved, the rest of the coat left shaggy) or a working-retriever style (entire coat clipped evenly to about an inch long). His tail is always shaved, leaving a plume on the end. He is black, brown, white, black-and-white, or brown-and-white.
3. His ears hang down but don't reach below his lower jaw. His eyes are black or brown. His nose is usually black, but is brown on brown dogs. His teeth meet in a scissors or level bite. His feet are webbed to aid in swimming.

Health Problems

He is susceptible to hip dysplasia, PRA, and GM-1 Storage Disease. Storage Disease is a fatal neurological disorder that evidences itself at six months of age by head bobbing and wobbliness. Buy only from OFA- and CERF-registered parents who have been tested for GM-1. Reputable breeders also test each puppy individually for Storage Disease before selling him.

Cautions when Buying

Don't choose the most independent PWD puppy and don't choose a timid puppy.

Rottweiler (ROTT-why-ler)

For Experienced Owners Only
Good with Children if Raised with Children
Large to Giant in Size
Short Coat

	High	Med	Low
EXERCISE REQUIRED		⬤	
TRIMMING/CLIPPING REQUIRED			⬤
AMOUNT OF SHEDDING		⬤	
ACTIVITY INDOORS			⬤
EASE OF TRAINING		⬤	
SOCIABILITY WITH STRANGERS			⬤

Temperament

He is calm, confident, steady tempered, and serious. Although quiet indoors, he's a robust dog who needs enough space and daily exercise that he's better in the suburbs or country. This is a powerful, purposeful, reserved breed who will protect his family, so he should be accustomed to people at an early age. He can be aggressive with other dogs. He knows his own mind, but he's very responsive to firm, fair obedience training. He must be taught not to pull on the leash. He may refuse commands from family members who have not established leadership over him. Don't play aggressive games like wrestling or tug-of-war. Ongoing socialization, exercise, companionship, and obedience training are necessities with a Rottweiler.

History

Developed in Rottweil, Germany, he drove the butcher's cattle to and from market. Legend has it that the butcher tied his profits to his dog's collar, where no highwayman would dare touch it. Today he is a guard dog, Schutzhund dog, home guardian, and companion. AKC popularity: 7th of 130 breeds.

Physical Features

1. Males stand 24–27 inches, females stand 22–25 inches. He weighs 85–115 pounds.
2. His short coat is coarse and dense and should be brushed once a week. He is black with rust markings above each eye, on his muzzle, throat, chest, legs and feet, and below his tail, and with narrow black streaks on his toes. A few white hairs are allowed on his chest.
3. His ears hang down. His eyes are brown with a fearless expression. His nose is black. His teeth meet in a scissors bite. His tail is docked.
4. Faults: His eyes should not be yellow or two different colors. The inside of his mouth should not be pink. His teeth should not meet in a level bite. His rust markings should not be light tan or spread too far.
5. Disqualifications: His teeth should not meet in an overshot or undershot bite. His coat should not be long. He should not have more than three missing teeth. He should not be any colors other than black-and-rust, such as solid black, or red and rust.

Health Problems

He is susceptible to hip dysplasia, bloat, retinal problems, and spinal-cord paralysis. Buy only from OFA-registered parents.

Cautions when Buying

THERE ARE MANY ROTTWEILERS AROUND— BE VERY CAREFUL. Look for a breeder who strives for good conformation and some working ability in obedience. Show-only breeders also produce good family pets, but Schutzhund breeders often produce dogs who are too aggressive. If you buy from a poor breeder or raise the dog incorrectly, you could wind up with an overly dominant, aggressive Rottie. Don't choose the boldest or most independent Rottweiler puppy or a timid puppy.

Longhaired Saint Bernard and Shorthaired Saint Bernard

These are varieties of the same breed. The Shorthaired is pictured.

For Experienced Owners Only
Good with Children
Giant in Size
Thick Medium-Length Coat or Short Coat

	High	Med	Low
EXERCISE REQUIRED		✦	
TRIMMING/CLIPPING REQUIRED			✦
AMOUNT OF SHEDDING	✦		
ACTIVITY INDOORS			✦
EASE OF TRAINING		✦	
SOCIABILITY WITH STRANGERS		✦	

Temperament

He is calm, sensible, gentle, and patient. Some are more outgoing, some are more introspective. He's quiet indoors, but he does best in the suburbs or country, preferably in a cold climate. He needs daily exercise whether he seems to want it or not. He also needs companionship and is not suited for the backyard. He's mildly friendly with strangers, but because he is a giant breed, he should be accustomed to people at an early age. He's usually fine with other animals. He wants to please and is responsive to patient obedience training that includes much praise and little jerking around; this powerful breed must be treated with respect and never hit. Don't play aggressive games like wrestling or tug-of-war. He drools heavily and wheezes.

History

At a hospice monastery near St. Bernard's Pass in the Swiss Alps, he was used by the monks as a rescue dog. On the trail, his sixth sense and keen hearing warned him from the paths of avalanches moments before they occurred. The Shorthaired Saint, the older of the two varieties, is far less common than is the Longhaired. The mountain hospice dogs were—and still are—shorthaired, because ice clings to longhaired coats. Today he is a companion, sometimes pulling sleds and carts. AKC popularity: 42nd of 130 breeds.

Physical Features

1. Males stand at least 27½ inches, females stand at least 25½ inches. He weighs 150–170 pounds.
2. The Shorthaired Saint has a short, dense coat. The Longhaired Saint has a longer, slightly wavy coat. Both varieties should be brushed once a week, but daily during shedding. He is red-and-white, brownish yellow-and-white, or brindle-and-white.
3. His ears hang down. His eyes are dark brown with a friendly expression. His nose is black. His teeth meet in a scissors, level, or undershot bite.

Health Problems

He is a short-lived breed (ten years) and susceptible to hip dysplasia, bloat, tumors, skin conditions, heart disease, and eyelid abnormalities. Buy only from OFA-registered parents.

Cautions when Buying

BE CAREFUL. This huge, powerful breed has been shamelessly exploited by poor breeders, and if you buy from them or raise the dog incorrectly, you could end up with a sickly, dominant, or aggressive Saint Bernard. Don't choose a bold Saint puppy or a timid puppy.

Samoyed (SAM-oy-ed)

For Experienced Owners Only
Good with Children
Medium to Large in Size
Thick Medium-Length Coat

	High	Med	Low
EXERCISE REQUIRED		✦	
TRIMMING/CLIPPING REQUIRED			✦
AMOUNT OF SHEDDING	✦		
ACTIVITY INDOORS	✦		
EASE OF TRAINING			✦
SOCIABILITY WITH STRANGERS		✦	

Temperament

He is gentle, dependable, and very playful. He can adapt to the city if given companionship and vigorous exercise, but he loves the outdoors, especially in the cold. Too much solitary confinement makes him restless and very destructive—he is a notorious chewer. He's usually mildly friendly with strangers and good with other animals. One of the brightest of the sled-dog breeds, yet still obstinate and independent, he needs early and firm obedience training. The commands *heel*, *down*, and *come* are especially important in keeping him from becoming too boisterous. He has a jolly sense of humor and often exhibits it when disobeying. His black lips curve upward so that he always looks like he's smiling. He can be hard to housebreak. He barks a lot in a high, piercing voice.

History

An old Asian breed, he developed mostly on his own and was used by the nomadic Samoyed peoples of the Siberian tundra for herding and guarding their reindeer and for pulling their sleds. Today he is a companion who sometimes pulls carts and sleds. AKC popularity: 32nd of 130 breeds.

Physical Features

1. Males stand 21–23½ inches and weigh 55–70 pounds. Females stand 19–21 inches and weigh 45–55 pounds.
2. His coat is harsh with a ruff around his neck and longer hair on his chest, stomach, legs, and tail. He should be brushed twice a week ordinarily, but daily when he's shedding. He is white, cream, biscuit (beige), or white with biscuit shadings.
3. His ears prick up. His eyes are dark with an animated expression. His nose is black, brown, or flesh colored, and the color sometimes fades to pink in cold weather. His teeth meet in a scissors bite. His furry tail curls over his back.
4. Disqualifications: He should not be any color other than those listed, such as black. His eyes should not be blue.

Health Problems

He is susceptible to hip dysplasia, PRA, glaucoma, retinal detachment, and skin conditions. Buy only from OFA- and CERF-registered parents.

Cautions when Buying

Don't choose the boldest, noisiest, most energetic, or most independent Sammy puppy.

Siberian Husky

For Experienced Owners Only
Good with Children
Medium to Large in Size
Thick Medium-Length Coat

	High	Med	Low
EXERCISE REQUIRED		▰	
TRIMMING/CLIPPING REQUIRED			▰
AMOUNT OF SHEDDING	▰		
ACTIVITY INDOORS	▰		
EASE OF TRAINING			▰
SOCIABILITY WITH STRANGERS	▰		

Temperament

He is good-natured, demonstrative, and very playful. He's so adaptable that he does fine in the city if vigorously exercised, but he's energetic and loves the outdoors, so he's better in the suburbs or country, preferably in a cold climate. He is easily bored—too much confinement and too little to do makes him restless and destructive. He is a notorious chewer—some females will dig den holes in the sofa. He's friendly with strangers and usually fine with other family pets, but he needs careful outdoor supervision for he is an explorer with a wild background, and he may molest running cats, deer, and livestock. Don't let him off the leash unless in a safe, enclosed area or he could take off for parts unknown. His obstinate, independent nature calls for much early obedience training, but this is a clever breed, often yelping at the gentlest tug on the leash, then quickly seizing advantage of any indulgence. He can be mischievously disobedient and hard to housebreak. He howls a lot. Ongoing exercise, obedience training, and supervision are necessities with a Siberian.

History

He was developed by the *Chukchi* people of Siberia as a sled dog able to travel great distances with light loads at fast speeds. He also pulled Admiral Byrd's Antarctic expeditions. Today he's still a racing dog (unlike his cousin the Alaskan Malamute, the Siberian is built for speed rather than power) and a companion. AKC popularity: 20th of 130 breeds.

Physical Features

1. Males stand 21–23½ inches, females stand 20–22 inches. He weighs 35–60 pounds.
2. His coat is well furred with a dense undercoat. He should be brushed twice a week ordinarily, but brush daily when he's shedding. He is gray-and-white, black-and-white, red-and-white, or solid white.
3. His ears prick up. His eyes are brown, blue, or mixed. His nose is black on black and gray dogs; brown on red dogs; may be flesh colored on white dogs; and is sometimes pink-streaked. His teeth meet in a scissors bite. His furry tail is carried over his back but not tightly curled.
4. Disqualifications: He should not stand over the height specification.

Health Problems

He is susceptible to hip dysplasia, PRA, cataracts, corneal disorders, thyroid deficiency, and zinc-deficiency skin disease. Buy only from OFA- and CERF-registered parents.

Cautions when Buying

BE CAREFUL. If you buy this breed from a poor breeder, you could wind up with a tall, bony, hyperactive Siberian. Some breeders concentrate on racing dogs; their dogs usually have good temperaments, but some are too energetic to make good pets. Don't choose the boldest, noisiest, most rambunctious, or most independent Siberian puppy.

Standard Schnauzer

For Experienced Owners Only
Good with Older, Considerate Children
Medium in Size
Wiry Coat

	High	Med	Low
EXERCISE REQUIRED		🦴	
TRIMMING/CLIPPING REQUIRED	🦴		
AMOUNT OF SHEDDING			🦴
ACTIVITY INDOORS		🦴	
EASE OF TRAINING		🦴	
SOCIABILITY WITH STRANGERS			🦴

Temperament

He is bold, high-spirited, and reliable. Although lively indoors and out, he makes a fine city dog if exercised daily and allowed to participate in games and activities. He must have companionship and cannot be ignored. Reserved and protective, he should be accustomed to people at an early age. He can be aggressive with other dogs. He's extremely bright, but also clever, strong-willed, stubborn, and he has a fine sense of humor. He needs ongoing obedience training that is firm but kind and persuasive, because he will rebel against harshness and too much leash jerking. Never hit a Schnauzer or he could retaliate. He's very adaptable and makes an excellent traveling companion. He may refuse commands and guard objects from family members who have not established leadership over him. Don't play aggressive games like wrestling or tug-of-war.

History

All three breeds of Schnauzer (Miniature, Standard, and Giant) originated in Germany; they are separate breeds, the Standard being the oldest. He was a rat catcher, guardian of the marketplace produce stand, and during World War I a messenger carrier. Today he is a home guardian and companion. AKC popularity: 86th of 130 breeds.

Physical Features

1. Males stand 18½–19½ inches, females stand 17½–18½ inches. He weighs 30–45 pounds.
2. His coat is short and hard, with bushy eyebrows and a beard, and should be brushed and combed twice a week. He needs scissoring and coat shaping every three months. Coat shaping for pets is done by electric clipping. Coat shaping for show dogs is done by stripping (plucking) the dead hairs out one by one, because electric clipping softens the coat and fades the color. He is pepper-and-salt (dark gray or silvery gray with a liberal sprinkling of black and white hairs among the gray) or solid black (a white spot on his chest is allowed).
3. His ears may be cropped or left folded forward naturally. His eyes are dark brown. His nose is black. His teeth meet in a scissors or level bite. His tail is docked to 1 or 2 inches long.
4. Disqualifications: He should not stand more than ½ inch over or under the height specification.

Health Problems

He is a long-lived breed (fifteen years) and susceptible to hip dysplasia and cysts. Buy only from OFA-registered parents.

Cautions when Buying

Don't choose the boldest, most scrappy, most energetic, or most independent Standard Schnauzer puppy.

The Terrier Group

Airedale Terrier · American Staffordshire Terrier · Australian Terrier
Bedlington Terrier · Border Terrier · Bull Terrier · Cairn Terrier
Dandie Dinmont Terrier · Irish Terrier · Kerry Blue Terrier
Lakeland Terrier · Standard Manchester Terrier · Miniature Schnauzer
Norfolk Terrier · Norwich Terrier · Scottish Terrier · Sealyham Terrier
Skye Terrier · Smooth Fox Terrier · Soft-Coated Wheaten Terrier
Staffordshire Bull Terrier · Welsh Terrier · West Highland White Terrier
Wire Fox Terrier

Airedale Terrier

For Experienced Owners Only
Good with Older, Considerate Children
Medium to Large in Size
Wiry Coat

	High	Med	Low
EXERCISE REQUIRED		●	
TRIMMING/CLIPPING REQUIRED	●		
AMOUNT OF SHEDDING			●
ACTIVITY INDOORS		●	
EASE OF TRAINING		●	
SOCIABILITY WITH STRANGERS	●		

Temperament

He's more sensible than most terriers. A playful, rowdy handful as a puppy, he matures into a dignified, courageous adult. He's so adaptable that he does fine in the city if given enough daily exercise to work off his high energy and if allowed to participate in games and activities. He is sensible with strangers, but because he can be protective, he should be accustomed to people at an early age. He can be scrappy with other animals. Bold, stubborn, and headstrong, this capable dog needs early, consistent obedience training that is firm but fair. Never hit a terrier or he could retaliate. He may refuse commands from family members who have not established leadership over him. Don't play aggressive games like wrestling or tug-of-war.

History

He was developed in the English river valley of Aire from crosses of black-and-tan terriers and the shaggy Otter Hound. He hunted badgers, otters, and river rats. He was also a police dog and military messenger dog. Today he is a home guardian and companion, but he can still dispatch vermin. AKC popularity: 41st of 130 breeds.

Physical Features

1. He stands 22–23 inches and weighs 45–60 pounds.
2. His coat is hard and wiry, with bushy eyebrows and a beard, and needs to be brushed and combed twice a week. He needs scissoring and coat shaping every three months. For pets, coat shaping means electric clipping; for show dogs, it means stripping (plucking) the dead hairs out one by one, because electric clipping softens the coat and fades the color. His head, legs, and chest are tan, his back and sides are blackish (often mixed with red). A white blaze on his chest is allowed.
3. His ears fold forward. His eyes are dark with a keen expression. His nose is black. His teeth meet in a scissors or level bite. His tail is docked.
4. Faults: His eyes should not be yellow. His feet should not be white.

Health Problems

He is susceptible to hip dysplasia. Buy only from OFA-registered parents.

Cautions when Buying

Some breeders advertise "hunting and companion" Airedales—be sure these lines have good structure, health, and temperament before buying this type. Don't choose the boldest, most energetic, most independent Airedale puppy.

American Staffordshire Terrier

For Experienced Owners Only
Good with Older, Considerate Children
Medium to Large in Size
Smooth Coat

	High	Med	Low
EXERCISE REQUIRED		●	
TRIMMING/CLIPPING REQUIRED			●
AMOUNT OF SHEDDING		●	
ACTIVITY INDOORS	●		
EASE OF TRAINING			●
SOCIABILITY WITH STRANGERS		●	

Temperament

He is usually quiet and docile with an impressive, confident presence. He's fine in the city if vigorously exercised. He's also fine with strangers when his family is present (some are very friendly), but this is a powerful, protective dog who should be thoroughly accustomed to people at an early age. He can be very aggressive with other animals, although many AmStaffs will only attack other AmStaffs (or similar breeds). Still, he should not be allowed off the leash except in a safe, enclosed area. Although stubborn, he will respond to early, consistent obedience training that is firm but fair. He may refuse commands from family members who have not established leadership over him. Don't play aggressive games like wrestling or tug-of-war, and anyone who hits this breed is asking for trouble. He can be hard to housebreak. Lifelong socialization, exercise, training, and supervision are essential. All AmStaff owners must spend time and effort training their dogs, because every calm, obedient AmStaff seen on a public street can help counteract the current antidog hysteria sweeping the country.

History

He originated in England as a cross between fighting bulldogs and terriers. When he came to America, proponents of dogfighting developed their own strain called the American Pit Bull Terrier, which is recognized by the UKC but not the AKC. The American Staffordshire is owned by show fanciers who breed for docility, while the Pit Bull is often owned by underworld dogfighters and street criminals who breed for aggressiveness. The dogs making the headlines as vicious attackers are definitely *not* AKC American Staffordshire show dogs but UKC Pit Bulls, unregistered Pit Bulls, and Pit Bull mixes. The American Staffordshire is a home guardian and companion. AKC popularity: 77th of 130 breeds.

Physical Features

1. He stands 17–19 inches and weighs 50–80 pounds.
2. His hard coat needs only a quick brushing once a week. Colors include fawn, black, brindle (brownish with black stripes and flecks), patched, and others.
3. His ears may be cropped or left alone. His eyes are dark. His nose is black. His teeth meet in a scissors bite.
4. Faults: His eyes and nose should not be pink. He should not be black-and-tan, liver (brown), solid white, or mostly white.

Health Problems

He is susceptible to hip dysplasia and serious tumors. Buy only from OFA-registered parents.

Cautions when Buying

BE CAREFUL. This is a powerful breed, and if you buy from a poor breeder or raise the dog incorrectly, you could end up with an aggressive AmStaff. Be sure you're buying an AKC American Staffordshire Terrier, although UKC-registered American Pit Bull Terriers can be fine *if* they're from UKC show and obedience lines. *Don't buy from any other line.* Don't choose the boldest, most energetic, or most independent AmStaff puppy, and don't choose a timid puppy.

Australian Terrier 9

Fine for Novice Owners ✓
Good with Older, Considerate Children ✓
Little in Size ✓
Wiry Coat ✓

	High	Med	Low
EXERCISE REQUIRED ✓		✖	
TRIMMING/CLIPPING REQUIRED ✓		✖	
AMOUNT OF SHEDDING ✓			✖
ACTIVITY INDOORS ✓		✖	
EASE OF TRAINING ✓	✖		
SOCIABILITY WITH STRANGERS			✖

Temperament

He is one of the quietest, most obedient, and least demanding of the terriers but just as hardy, spirited, and plucky as the rest. He's so adaptable that he's easy to live with in any situation, but he does like outdoor walks. He's reserved with strangers and good with other animals. Quick to learn and usually eager to please, he responds very well to obedience training. He can snap if teased or provoked—never hit any terrier and don't play aggressive games like wrestling or tug-of-war. He likes to dig.

History

He was developed in Australia from various rough-coated terriers and once hunted snakes and rats. Today he is a companion, but many could still do a job on small vermin. AKC popularity: 78th of 130 breeds.

Physical Features

1. He stands about 10 inches and weighs 12–18 pounds.
2. His coat is short and harsh. Straggly hairs should be scissored and dead hairs stripped twice a year. Usually he is blue-black or silver-black, with a reddish-tan head and legs. He can also be sandy red (tan).
3. His ears prick up. His eyes are dark with a keen expression. His nose is black. His teeth meet in a scissors or level bite. His tail is docked.

Health Problems

He is a healthy breed, prone to no real problems.

Cautions when Buying

Don't choose a scrappy Aussie puppy or a timid puppy.

Bedlington Terrier

Fine for Novice Owners
Good with Older, Considerate Children
Small in Size
Curly Coat

	High	Med	Low
EXERCISE REQUIRED		⬥	
TRIMMING/CLIPPING REQUIRED	⬥		
AMOUNT OF SHEDDING			⬥
ACTIVITY INDOORS		⬥	⬥
EASE OF TRAINING			⬥
SOCIABILITY WITH STRANGERS		⬥	

Temperament

He is milder and less rowdy than some terriers, but despite his lamblike appearance, this little dog can be a fierce fighter. He can adapt to the city if given long walks and occasionally allowed to stretch out and gallop, but don't let him off the leash unless in a safe, enclosed area—he is a quick, agile chaser. Some are friendly with strangers, some are reserved, some are timid or nervous, so he should be accustomed to people and noises at an early age. He can be scrappy with strange dogs, jealous of other pets, and can chase small animals like cats and rabbits. Stubborn and headstrong, he needs firm obedience training, but it must include more praise and encouragement than leash-jerking. He can suddenly snap with lightning reflexes if teased or provoked. Never hit a terrier, and don't play aggressive games like wrestling or tug-of-war. He barks a lot and can be hard to housebreak. He likes to dig.

History

His name comes from Bedlington, England, where he was developed by miners as a tough hunter of badgers and rats. Occasionally he was used as a dogfighter and only later did he become a fashionable pet. Today he is a companion, but he can still dispatch vermin. AKC popularity: 108th of 130 breeds.

Physical Features

1. He stands 15–17½ inches and weighs 17–23 pounds.
2. His coat is a short curly mixture of hard and soft hair that should be brushed twice a week. He needs scissoring and expert coat shaping every two months. He is blue (actually whitish-blue), sandy (tan), or liver (brown), sometimes with tan markings over his eyes, on his legs and chest, and under his tail. A puppy's coat lightens to his adult color by one year of age.
3. His ears hang down and have a silky tassel at each tip. His eyes are brown or hazel with a mild expression. His nose is black on blue dogs, brown on sandies and livers. His teeth meet in a scissors or level bite.

Health Problems

He is susceptible to PRA, cataracts, retinal dysplasia, kidney disease, thyroid deficiency, and copper toxicosis (a severe inherited liver disease). Buy only from liver-diagnosed parents.

Cautions when Buying

Don't choose the most scrappy or independent Bedlington puppy, and don't choose a timid puppy.

Border Terrier ¹⁰

Fine for Novice Owners ✓
Good with Children ✓
Small in Size ✓
Wiry Coat ✓

	High	Med	Low
EXERCISE REQUIRED ✓		🦴	
TRIMMING/CLIPPING REQUIRED ✓		🦴	
AMOUNT OF SHEDDING ✓			🦴
ACTIVITY INDOORS ✓		🦴	
EASE OF TRAINING ✓	🦴		
SOCIABILITY WITH STRANGERS ✓	🦴		

Temperament

He is quieter, milder-mannered, and more obedient than most terriers, but still very hardy and plucky. Some Borders are tougher and more work-oriented than others. He's so adaptable that he's fine in any home that gives him daily companionship, long walks, and an occasional run in a safe, enclosed area. He's usually friendly with strangers, but some are timid, so he should be accustomed to people and noises at an early age. Unlike many terriers, he's usually fine with other dogs, but must be watched around small pets like cats and rabbits, for he can be a single-minded chaser. Borders are inquisitive and explorative, and can get themselves into trouble by crawling into tight holes; although their loose skin often enables them to wriggle out by themselves, you should inspect your home for openings that a Border may not be able to resist. Eager to please and sensitive to correction, he's very responsive to obedience training. Never hit a terrier, and don't play aggressive games like wrestling or tug-of-war. He likes to dig; some will bark a bit.

History

This hardy hunter of fox, badger, and barnyard varmints originated in the rugged Cheviot Hills on the border between England and Scotland. Today he is a companion, but he can still dispatch vermin and makes an excellent farm dog. AKC popularity: 88th of 130 breeds.

Physical Features

1. He stands 12–15 inches and usually weighs 15–18 pounds.
2. His coat is short and harsh and should be brushed twice a week. Straggly hairs should be scissored and dead hairs stripped twice a year. He is wheaten (tan) or reddish brown, with black-tipped hairs and dark ears. Blue-and-tan is less common. A little white on his chest is allowed.
3. His ears hang down. His eyes are dark with a keen expression. His nose is black. His teeth meet in a scissors bite.
4. Faults: He should not have white hairs on his feet.

Health Problems

He is susceptible to some congenital heart problems. Like many stoic terriers, he is relatively insensitive to pain and shows few signs of illness; it is up to his owner to watch his health carefully.

Cautions when Buying

Don't choose the most scrappy or independent Border puppy, and don't choose a timid puppy.

Bull Terrier

For Experienced Owners Only
Good with Older, Considerate Children
Medium in Size
Smooth Coat

	High	Med	Low
EXERCISE REQUIRED		▰	
TRIMMING/CLIPPING REQUIRED			▰
AMOUNT OF SHEDDING		▰	
ACTIVITY INDOORS		▰	▰
EASE OF TRAINING			▰
SOCIABILITY WITH STRANGERS		▰	

Temperament

The "White Cavalier" is a sweet-tempered gentleman. He's energetic, assertive, and playful but fine in the city if exercised and made a part of the family. He cannot be owned and ignored, for he is easily bored and can be mischievous. Some are calmer and more laid-back than others. Usually he's friendly with strangers, but he should be accustomed to people at an early age. He can be very aggressive with other dogs, so don't let him off the leash except in a safe, enclosed area. Some should not be kept around small pets like cats and rabbits. He's very stubborn and needs patient obedience training that includes much praise and little jerking around. This is a funny, witty breed; he likes to invent new rules and has a mind of his own. Don't play aggressive games like wrestling or tug-of-war.

History

Developed in England from crosses of bulldogs and an extinct white terrier, he was a dogfighter, but this breed is *emphatically not* the Pit Bull Terrier that has been making the headlines. (See the American Staffordshire Terrier profile for more information.) The ferocity has been bred out of this English Bull Terrier with the large prick ears and long face. He is a home guardian and companion. AKC popularity: 65th of 130 breeds.

Physical Features

1. He stands 15–22 inches and weighs 35–80 pounds.
2. His hard coat needs only a quick brushing once a week. He comes in two color varieties: *White* (either solid white or mostly white with a few colored patches on his head) and *Colored* (preferably brindle, black-and-tan, or fawn, with white markings allowed on his head, chest, neck, feet, and tail tip).
3. His ears prick up. His eyes are dark and triangular with a keen expression. His nose is black—a pink nose on a white puppy usually darkens later. His teeth meet in a scissors or level bite.
4. Disqualifications: His eyes should not be blue. A colored Bull Terrier should not be mostly white.

Health Problems

He is susceptible to inherited deafness, patella problems, and skin problems.

Cautions when Buying

BE CAREFUL. This is a powerful, independent breed, and if you buy from a poor breeder or raise the dog incorrectly, you could end up with an overly dominant, aggressive Bull Terrier. Don't choose the boldest, most boisterous, or most independent Bullie puppy, and don't choose a timid puppy.

Cairn Terrier

Fine for Novice Owners ✓
Good with Older, Considerate Children ✓
Little in Size ✓
Wiry Coat ✓

	High	Med	Low
EXERCISE REQUIRED ✓		🦴	
TRIMMING/CLIPPING REQUIRED ✓	🦴		
AMOUNT OF SHEDDING ✓			🦴
ACTIVITY INDOORS	🦴		
EASE OF TRAINING ✓		🦴	
SOCIABILITY WITH STRANGERS		🦴	

Temperament

He is the breed many people picture when they hear "terrier." He is hardy, spirited, plucky, and bold. He likes his walks but is adaptable to any home in which he can be a busybody and a full participant. He is reserved with strangers, scrappy with other animals. Don't let him off the leash except in a safe, enclosed area, for he is an independent chaser. Assertive but cheerful, with typical terrier stubbornness, he does want to please and responds well to firm, fair obedience training. To come when called is a very important command for him to learn. Never hit a terrier and don't play aggressive games like wrestling or tug-of-war. He can guard his food and toys from family members who have not established leadership over him. He is a digger, and some bark a bit.

History

He was developed in Scotland as a rugged hunter of fox, otter, and rats, and he takes his name from the rocky cliff dens (cairns) where these animals took refuge. Today he is a companion, but he can still dispatch vermin and does well on a farm. AKC popularity: 33rd of 130 breeds.

Physical Features

1. He stands about 10 inches and weighs 13–14 pounds.
2. His coat is harsh and should be brushed and combed twice a week. Straggly hairs should be scissored and dead hairs stripped twice a year. Colors include wheaten (beige), red, and many shades of brindle (brown, gray, or silver, with black stripes and flecks). A puppy's coat may change color as he grows.
3. His ears prick up. His eyes are hazel with a keen expression. His nose is black. His teeth meet in a scissors or level bite.
4. Faults: His eyes should not be yellow. His nose should not be light colored or flesh colored. He should not have any white patches or a white foot.

Health Problems

He is susceptible to skin allergies.

Cautions when Buying

Don't choose the boldest, most scrappy, or most independent Cairn puppy.

Dandie Dinmont Terrier

Fine for Novice Owners ✓
Good with Older, Considerate Children ✓
Small in Size ✓
Unusual Coat (see Physical Features 2. below) ✓

	High	Med	Low
EXERCISE REQUIRED ✓		✦	
TRIMMING/CLIPPING REQUIRED ✓	✦		
AMOUNT OF SHEDDING		✦	
ACTIVITY INDOORS ✓		✦	
EASE OF TRAINING			✦
SOCIABILITY WITH STRANGERS			✦

Temperament

This hardy, plucky little dog is one of the brightest of the terriers but also one of the most independent. Some are one-person dogs. With some exercise, he's fine in the city. He's reserved with strangers, scrappy with other animals. Assertive and strong willed, with a definite mind of his own, he does respond to firm, fair obedience training. Never hit a terrier, and don't play aggressive games like wrestling or tug-of-war. He can guard his food and toys from family members who have not established leadership over him. He can be hard to housebreak and he is a digger. Although some are very quiet, some can be barkers, and he has a surprisingly deep bark for such a small dog.

History

He was developed in Scotland as a rugged hunter of otter and badger. In Sir Walter Scott's novel about the farmer Dandie Dinmont, tough hunting terriers were described, and these terriers became known as Dandie Dinmont's dogs. The breed was once favored by gypsies and later became fashionable with high society. Today he is a companion, but he can still dispatch vermin. AKC popularity: 119th of 130 breeds.

Physical Features

1. He stands 8–11 inches and weighs 18–24 pounds.
2. His coat is about 2 inches long, a unique mixture of hard and soft hair, with a bushy topknot. He should be brushed and combed twice a week. He needs scissoring and coat shaping every three months. For pets, coat shaping means electric clipping; for show dogs, it means stripping (plucking) the dead hairs out one by one, because electric clipping softens the coat and fades the color. A puppy born black-and-tan will turn "pepper-colored" (blue-black to silvery gray). A puppy born brownish with a black mask will turn "mustard-colored" (reddish brown to fawn-tan). These color changes take place by about eight months. His head and topknot are always creamy white. Some white on his chest or toenails is okay.
3. His ears hang down with a pompom on each end. His eyes are dark hazel with a wise, determined expression. His nose is dark. His teeth meet in a scissors bite. His front legs are shorter than his hind legs so that he seems to walk downhill.

Health Problems

He is susceptible to spinal disc problems—don't let him jump off high furniture.

Cautions when Buying

Don't choose the boldest, most scrappy, or most independent Dandie puppy.

Irish Terrier

Fine for Novice Owners ✓
Good with Children ✓
Medium in Size
Wiry Coat ✓

	High	Med	Low
EXERCISE REQUIRED ✓		●	
TRIMMING/CLIPPING REQUIRED	●		
AMOUNT OF SHEDDING ✓			●
ACTIVITY INDOORS ✓		●	
EASE OF TRAINING			●
SOCIABILITY WITH STRANGERS		●	

Temperament

This is one of the hardiest, pluckiest, boldest, and most spirited of the terriers. He is also very playful. He is both a gallant gentleman and a fearless daredevil who charges headlong into any situation. He can adapt to any home with daily exercise and companionship. He thrives on games and activities. Don't let him off the leash except in a safe, enclosed area—he is an explorer, an independent chaser, and very fast. He's sensibly friendly with strangers but very protective, so he should be accustomed to people at an early age. He's scrappy with other animals. Although his great stubbornness and energy call for ongoing obedience training (*come* and *stay* are especially important commands), he can learn anything—this is one of the most capable and versatile of all breeds. Never hit any terrier, and don't play aggressive games like wrestling or tug-of-war. He can guard his food and toys from family members who have not established leadership over him. He is a digger.

History

Developed in Ireland, he was a true working breed: farm dog, vermin hunter, water retriever, big-game hunter, messenger dog, and patrol dog. Today he is a home guardian and companion, but he can still dispatch vermin. AKC popularity: 107th of 130 breeds.

Physical Features

1. He stands about 18 inches and weighs 25–27 pounds.
2. His coat is short and wiry, with bushy eyebrows and a beard, and should be brushed and combed twice a week. He needs much scissoring and coat shaping every three months. For pets, coat shaping means electric clipping; for show dogs, it means stripping (plucking) the dead hairs out one by one, because electric clipping softens the coat and fades the color. He is usually red or golden red; wheaten is less common. Some puppies are born with a few black hairs that usually disappear as they grow. A little white on his chest is allowed.
3. His ears fold forward. His eyes are dark with a fiery expression. His nose is black. His teeth meet in a scissors or level bite. His tail is docked.
4. Faults: His eyes should not be yellow. His ears should not hang sideways. He should not have any white hairs, except on his chest.

Health Problems

He is healthy, prone to no real problems.

Cautions when Buying

Don't choose the boldest, most independent, or most energetic Irish puppy.

Kerry Blue Terrier

For Experienced Owners Only
Good with Older, Considerate Children
Medium in Size
Unusual Coat (See Physical Features 2. below)

	High	Med	Low
EXERCISE REQUIRED		●	
TRIMMING/CLIPPING REQUIRED	●		
AMOUNT OF SHEDDING			●
ACTIVITY INDOORS		●	
EASE OF TRAINING			●
SOCIABILITY WITH STRANGERS			●

Temperament

Like the Irish Terrier, this hardy, spirited breed is extraordinarily capable and versatile, but he's moodier and more fickle—it can be difficult to read his thoughts or predict his actions. He's highly adaptable to any home if given daily exercise and companionship. He loves games and activities. Reserved and protective with strangers, he should be accustomed to people at an early age. He is scrappy with other animals. Don't let him off the leash except in a safe, enclosed area—he is an explorer, an independent chaser, and very fast. His great stubbornness and energy call for firm, ongoing obedience training (*come* and *stay* are especially important commands). This breed will take clever advantage of those who indulge him. Never hit a terrier, and don't play aggressive games like wrestling or tug-of-war. He can refuse commands from family members who have not established leadership over him. He can be quick to bite if startled or teased. He tends to bark and dig.

History

This capable breed originated in County Kerry, Ireland, as a small-game hunter, water retriever, sheepherder, and general farm dog. He was even trained as a police dog. Today he is a home guardian and companion, but he can still dispatch vermin. AKC popularity: 89th of 130 breeds.

Physical Features

1. He stands 17–20 inches and weighs 30–40 pounds.
2. His coat is silky and wavy with much facial hair. He should be brushed and combed twice a week. He needs scissoring and coat shaping about every three months. A puppy is born black, his coat going through gradual color phases (including blackish brown) before settling on steel blue or silvery gray at around eighteen months. His head, feet, and tail may stay black.
3. His ears fold forward. His eyes are dark with a keen expression. His nose is black. His teeth meet in a scissors or level bite. His tail is docked.
4. Disqualifications: He should not have dewclaws on his hind legs. He should not be solid black.

Health Problems

He is a long-lived breed (fifteen years) but susceptible to eye problems, cysts, and tumors.

Cautions when Buying

Don't choose the boldest, most independent, or most energetic Kerry Blue puppy.

Lakeland Terrier

Fine for Novice Owners ✓
Good with Older, Considerate Children ✓
Small in Size ✓
Wiry Coat ✓

	High	Med	Low
EXERCISE REQUIRED ✓		🦴	
TRIMMING/CLIPPING REQUIRED ✓	🦴		
AMOUNT OF SHEDDING ✓			🦴
ACTIVITY INDOORS	🦴		
EASE OF TRAINING			🦴
SOCIABILITY WITH STRANGERS			🦴

Temperament

Although not as wildly boisterous as some terriers, he's still feisty and spirited. He has stronger working instincts than many terriers, but he's fine in the city if exercised, played with, and given companionship. He's usually reserved with strangers and can be scrappy with other animals. Don't let him off the leash except in a safe, enclosed area, for he is a fast, independent chaser. His great stubbornness calls for firm obedience training, but he is sensitive—don't jerk him endlessly. Never hit a terrier, and don't play aggressive games like wrestling or tug-of-war. He can guard his food and toys from family members who have not established leadership over him. He likes to dig. Some are quiet, some are barkers.

History

He originated in the Lake District of England, and was used with Foxhounds to hunt foxes and with Otterhounds to hunt otters. Today he is a companion, but he can still dispatch vermin. AKC popularity: 99th of 130 breeds.

Physical Features

1. He stands 13–15 inches and weighs about 17 pounds.
2. His coat is short and wiry, with bushy eyebrows and a beard, and should be brushed and combed twice a week. He needs scissoring and coat shaping every three months. For pets, coat shaping means electric clipping; for show dogs, it means stripping (plucking) the dead hairs out one by one, because electric clipping softens the coat and fades the color. A puppy is born dark but lightens to reddish or wheaten (often with black-tipped hairs) by adulthood. Less common are blue, blue-and-tan, black, black-and-tan, and liver (brown).
3. His ears fold forward. His eyes are black or brown with an alert expression. His nose is black on most colors, but brown on liver dogs. His teeth meet in a scissors or level bite. His tail is docked.
4. Disqualifications: His teeth should not meet in an overshot or undershot bite.

Health Problems

He is a healthy breed, prone to no real problems.

Cautions when Buying

Don't choose the boldest, scrappiest, most independent Lakeland puppy.

Standard Manchester Terrier

Fine for Novice Owners
Good with Older, Considerate Children
Small in Size
Smooth Coat

	High	Med	Low
EXERCISE REQUIRED		▪	
TRIMMING/CLIPPING REQUIRED			▪
AMOUNT OF SHEDDING		▪	
ACTIVITY INDOORS	▪		
EASE OF TRAINING		▪	
SOCIABILITY WITH STRANGERS			▪

Temperament

This hardy, lively little dog is cleaner, more responsive, and better mannered than some terriers. He's fine in the city if exercised and played with, but don't let him off the leash except in a safe, enclosed area, for he is an independent chaser. He's reserved with strangers, scrappy with other animals. Obedience training must be firm, for he has a mind of his own, but he is also sensitive, so don't jerk him around. Never hit a terrier, and don't play aggressive games like wrestling or tug-of-war. He can guard his food and toys from family members who have not established leadership over him. He can suddenly snap if startled or annoyed. He likes to dig.

History

He was developed in Manchester, England, as a cross of Whippets and tough hunting terriers. There are two varieties: the Standard Manchester, which is a member of the AKC Terrier Group, and the Toy Manchester, which is a member of the Toy Group. The Standard is a companion, but he can still dispatch vermin. AKC popularity: 84th of 130 breeds, but this includes both Standards and Toys.

Physical Features

1. He stands 15–17 inches and weighs 12–22 pounds.
2. His hard coat needs only a quick brushing once a week. He is black with rich tan markings above his eyes, on his cheeks, throat, legs and feet, and under his tail. There is a black spot called a "thumb mark" on each front ankle, and narrow black stripes on all toes.
3. His ears may be cropped, naturally pricked up, or left folded forward. His eyes are dark with a keen expression. His nose is black. His teeth meet in a scissors or level bite.
4. Disqualifications: He should not be any color other than black-and-tan. He should not have a white stripe or patch measuring ½ inch or more. He should not weigh over 22 pounds.

Health Problems

He is susceptible to glaucoma and bleeding disorders.

Cautions when Buying

Don't choose the most independent or most scrappy Manchester puppy, and don't choose a timid puppy.

Miniature Schnauzer

Fine for Novice Owners
Good with Older, Considerate Children
Small in Size
Wiry Coat

	High	Med	Low
EXERCISE REQUIRED		🦴	
TRIMMING/CLIPPING REQUIRED	🦴		
AMOUNT OF SHEDDING			🦴
ACTIVITY INDOORS	🦴	🦴	
EASE OF TRAINING		🦴	
SOCIABILITY WITH STRANGERS		🦴	

Temperament

This plucky, playful little dog is more pleasant and more obedient than many terriers. He likes his walks but is fine in any home so long as he can participate in games and activities. Some are friendly with strangers, some are reserved, some are a bit timid. He's usually fine with other family pets—although he may chase the family cat for fun, he's seldom serious about it. Some can be a bit scrappy with strange dogs. Although he knows his own mind and often displays an obstinate resistance to walking on the leash, he is very bright and responds well to obedience training that is not too harsh. Never hit a terrier, and don't play aggressive games like wrestling or tug-of-war. He likes to be spoiled and doesn't take as much advantage of indulgence as some breeds. He's very adaptable and makes an excellent traveling companion. He can guard his food and toys from family members who have not established leadership over him. He is a barker and a digger.

History

All three Schnauzers (Miniature, Standard, and Giant) originated in Germany and are separate breeds. The Miniature was originally a vermin hunter but today is a companion, the most popular of all the terriers. AKC popularity: 11th of 130 breeds.

Physical Features

1. He stands 12–14 inches and weighs 13–15 pounds.
2. His coat is short and wiry, with bushy eyebrows and a beard, and should be brushed and combed twice a week. He needs scissoring and coat shaping every three months. For pets, coat shaping means electric clipping; for show dogs, it means stripping (plucking) the dead hairs out one by one, because electric clipping softens the coat and fades the color. He's usually salt-and-pepper (gray with a liberal sprinkling of black and white hairs, with silvery muzzle, chest, and legs, and sometimes with tan shadings). He can also be solid black (a white spot on his chest is allowed) or black-and-silver.
3. His ears may be cropped or left folded forward naturally. His eyes are dark with a keen expression. His nose is black. His teeth meet in a scissors bite. His tail is docked.
4. Disqualifications: He should not stand over or under the height specification. He should not be solid white or have any white patches on his body.

Health Problems

He is susceptible to cataracts and corneal disorders, kidney stones, bleeding disorders, liver disease, heart disease, diabetes, skin conditions, and cysts.

Cautions when Buying

THERE ARE MANY MINIATURE SCHNAUZERS AROUND—BE VERY CAREFUL. If you buy this breed from a poor breeder, you could wind up with a sickly, noisy, snappy Mini Schnauzer. Don't choose the most scrappy or most independent Mini Schnauzer puppy, and don't choose a timid puppy.

Norfolk Terrier

Fine for Novice Owners ✓
Good with Older, Considerate Children ✓
Little in Size ✓
Wiry Coat ✓

	High	Med	Low
EXERCISE REQUIRED ✓		●	
TRIMMING/CLIPPING REQUIRED ✓		●	
AMOUNT OF SHEDDING ✓			●
ACTIVITY INDOORS	●		
EASE OF TRAINING ✓		●	
SOCIABILITY WITH STRANGERS		●	

Temperament

He is a true representative of what a terrier is supposed to be: hardy, spirited, full of fire, assertive, but thoroughly companionable. He can adapt to any home with outdoor exercise. He's very sociable, very busy, and demands full participation in all activities. He's especially suited to a young athletic person or couple. Usually he's friendly with strangers and less scrappy with other dogs than many terriers. Don't let him off the leash except in a safe, enclosed area, for he is a fast, independent chaser. Many terrier enthusiasts feel that the Norfolk is much "softer" and less obstinate than the Norwich and must be trained very gently, with much praise and no leash-jerking. Never hit a terrier, and don't play aggressive games like wrestling or tug-of-war. Puppies should be thoroughly socialized, for some individuals can be a bit timid. He can be hard to housebreak and he is a tireless digger and barker.

History

Originating in England as a barnyard ratter, he became a fad with Cambridge University students. Near the city of Norwich, he was developed into a hunt terrier, riding in the saddlebag as the huntsmen-on-horseback followed the foxhounds. The hunt terrier was dropped to the ground when the fox went to ground, as it was his job to bolt the fox from his den. Originally this was a single breed called the Norwich, but recently it was split into the Norfolk with hanging ears and the Norwich with prick ears. Today he is a companion, but can still dispatch vermin and makes an excellent farm dog. AKC popularity: 104th of 130 breeds.

Physical Features

1. He stands 9–10 inches and weighs 11–12 pounds.
2. His coat is short and harsh and needs brushing and combing twice a week, and stripping of dead hairs twice a year. He's usually reddish or wheaten, sometimes with black-tipped hairs. Black-and-tan is less common.
3. His ears hang down. His eyes are dark with a keen expression. His nose is black. His teeth meet in a scissors bite. His tail is docked.

Health Problems

He is a healthy breed, prone to no real problems.

Cautions when Buying

Don't choose the boldest, most independent, or most energetic Norfolk puppy.

Norwich Terrier 8

Fine for Novice Owners ✓
Good with Older, Considerate Children ✓
Little in Size ✓
Wiry Coat ✓

	High	Med	Low
EXERCISE REQUIRED ✓		▪	
TRIMMING/CLIPPING REQUIRED ✓		▪	
AMOUNT OF SHEDDING ✓			▪
ACTIVITY INDOORS	▪		
EASE OF TRAINING ✓		▪	
SOCIABILITY WITH STRANGERS		▪	

Temperament

He is a true representative of what a terrier is supposed to be: hardy, spirited, full of fire, assertive, but thoroughly companionable. He can adapt to any home with outdoor exercise. He's very sociable, very busy, and demands full participation in all activities. He's especially suited to a young athletic person or couple. Usually he's friendly with strangers and less scrappy with other dogs than many terriers. Don't let him off the leash except in a safe, enclosed area, for he is a fast, independent chaser. Although obstinate and slow to mature, he responds fairly well to firm, fair obedience training. Never hit a terrier, and don't play aggressive games like wrestling or tug-of-war. Puppies should be thoroughly socialized, for some individuals can be a bit timid. He can be hard to housebreak and he is a tireless digger and barker.

History

Originating in England as a barnyard ratter, he became a fad with Cambridge University students. Near the city of Norwich, he was developed into a hunt terrier, riding in the saddlebag as the huntsmen-on-horseback followed the foxhounds. The hunt terrier was dropped to the ground when the fox went to ground, as it was his job to bolt the fox from his den. Originally this was a single breed called the Norwich, but recently it was split into the Norfolk with hanging ears and the Norwich with prick ears. Today he is a companion, but can still dispatch vermin and makes an excellent farm dog. AKC popularity: 101st of 130 breeds.

Physical Features

1. He stands 9–10 inches and weighs 11–12 pounds.
2. His coat is short and harsh, and needs brushing and combing twice a week, and stripping of dead hairs twice a year. He's usually reddish or wheaten, sometimes with black-tipped hairs. Black-and-tan is less common.
3. His ears prick up. His eyes are dark with a keen expression. His nose is black. His teeth meet in a scissors bite. His tail is docked.

Health Problems

He is a healthy breed, prone to no real problems.

Cautions when Buying

Don't choose the boldest, most independent, or most energetic Norwich puppy.

Scottish Terrier

Fine for Novice Owners ✓
Good with Older, Considerate Children ✓
Small in Size ✓
Wiry Coat ✓

	High	Med	Low
EXERCISE REQUIRED ✓		▰	
TRIMMING/CLIPPING REQUIRED ✓	▰		
AMOUNT OF SHEDDING ✓			▰
ACTIVITY INDOORS ✓		▰	
EASE OF TRAINING			▰
SOCIABILITY WITH STRANGERS			▰

Temperament

Playful and friendly as a puppy, he matures into a dignified adult—moodier and more independent than most terriers. He has been called staunchly self-reliant, self-assured, and fearless—also dour and crusty. Many are one-person dogs, very sensitive to their owner's moods. He's fine for the city if taken out for walks and played with. Reserved with strangers, he should be accustomed to people at an early age and not allowed to be sharp. He can be scrappy with other animals. He's stubborn minded, but also sensitive to correction, so obedience training must be consistent but persuasive—don't push this breed too far. Never hit a terrier, and don't play aggressive games like wrestling or tug-of-war. He makes an excellent traveling companion. He can challenge family members who have not established leadership over him. He is a barker and a digger.

History

He was developed in the rocky Scottish Highlands as a hunter of fox and vermin, and was originally called the Aberdeen Terrier. Today he is a companion, but he can still dispatch vermin. AKC popularity: 31st of 130 breeds.

Physical Features

1. He stands about 10 inches and weighs 18–22 pounds.
2. His coat is about 2 inches long and wiry, with bushy eyebrows and a beard, and should be brushed and combed twice a week. He needs scissoring and coat shaping every three months. For pets, coat shaping means electric clipping; for show dogs, it means stripping (plucking) the dead hairs out one by one, because electric clipping softens the coat and fades the color. He is black, dark gray, brindle (shades of brown, red, gold, or silver, with black stripes and flecks), sandy, or wheaten. A little white on his chest is allowed.
3. His ears prick up. His eyes are dark with a keen expression. His nose is black. His teeth meet in a scissors or level bite.

Health Problems

He is susceptible to von Willebrand's Disease, Scottie cramp (a minor condition where the dog has some difficulty walking), skin problems, flea allergies, and jawbone disorders.

Cautions when Buying

Don't choose the boldest, scrappiest, or most independent Scottie puppy.

Sealyham Terrier

Fine for Novice Owners ✓
Good with Older, Considerate Children ✓
Small in Size ✓
Wiry Coat ✓

	High	Med	Low
EXERCISE REQUIRED ✓		✖	
TRIMMING/CLIPPING REQUIRED ✓	✖		
AMOUNT OF SHEDDING ✓			✖
ACTIVITY INDOORS ✓			✖
EASE OF TRAINING			✖
SOCIABILITY WITH STRANGERS			✖

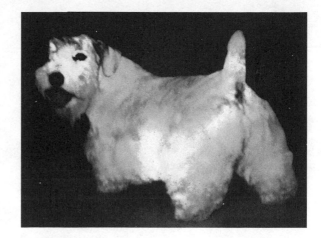

Temperament

This hardy dog is not as boisterous as some terriers, but he is more independent, more determined, and more self-willed. One of the few terriers who is calm indoors, he's fine in the city, but he needs long walks and an occasional romp in a safe, enclosed area. He is reserved with strangers and can be scrappy with other animals. His great stubbornness calls for early, firm obedience training, but as with all terriers, never hit him, jerk him around, or play aggressive games like wrestling or tug-of-war. He often exhibits a sense of humor when disobeying. He can guard his food and toys from family members who have not established leadership over him. He can be a barker and has a surprisingly deep bark for such a small dog. He likes to dig.

History

Developed in Wales, mostly by Captain John Edwardes, he was named after the Sealy River. He was a rugged hunter of fox, badger, weasel, and otter. Today he is a companion, but he can still dispatch vermin. He is the rarest of all the terriers. AKC popularity: 120th of 130 breeds.

Physical Features

1. He stands about 10½ inches and weighs 22–24 pounds.
2. His coat is hard and wiry, with bushy eyebrows and a beard, and should be brushed and combed twice a week. He needs scissoring and coat shaping every three months. For pets, coat shaping means electric clipping; for show dogs, it means stripping (plucking) the dead hairs out one by one, because electric clipping softens the coat and fades the color. He is white, with a few lemon, gray, or tannish markings on his head and ears.
3. His ears hang down. His eyes are dark with a keen expression. His nose is black. His teeth meet in a scissors or level bite. His tail is docked.
4. Faults: His nose should not be white, red, or spotted. His ears should not prick up completely or halfway.

Health Problems

He is a long-lived breed (fifteen years) but susceptible to skin conditions, back problems, glaucoma, and deafness. Check puppies' and parents' hearing by clapping and whistling.

Cautions when Buying

Don't choose the boldest, scrappiest, or most independent Sealy puppy.

Skye Terrier

For Experienced Owners Only
Good with Older, Considerate Children
Small in Size
Long Coat

	High	Med	Low
EXERCISE REQUIRED		✦	
TRIMMING/CLIPPING REQUIRED			✦
AMOUNT OF SHEDDING		✦	
ACTIVITY INDOORS			✦
EASE OF TRAINING			✦
SOCIABILITY WITH STRANGERS			✦

Temperament

He is hardy, spirited, plucky, and bold, but more serious and introspective than most terriers. He is keenly sensitive to moods and needs a lot of personal attention from his owner—he cannot be ignored. He likes to be included and needs to feel wanted. One of the few terriers who is calm indoors, he does fine in the city if given long walks and an occasional romp in a safe, enclosed area. He is reserved with strangers and should be accustomed to people at an early age. He needs careful supervision around other animals, for he is a fearless, agile chaser with lightning reflexes. He's mildly stubborn but does respond to firm, fair, patient obedience training. Never hit a terrier, and don't play aggressive games like wrestling or tug-of-war. He may bite if harshly disciplined, suddenly touched, or teased. He likes to dig.

History

One of the oldest of the terriers, he originated on the rugged Isle of Skye off the coast of Scotland as a tough vermin hunter. When his flowing coat became fashionable in the courts of London, he was cherished by the nobility. The legendary "Greyfriars Bobby," owned by a pauper shepherd, was so faithful that he daily lay beside his master's grave until his own death ten years after. The motto of the Skye Club of Scotland fits this breed perfectly: "Wha daur meddle wi' me." Today he is a companion, but he can still dispatch vermin. AKC popularity: 110th of 130 breeds.

Physical Features

1. He stands about 10 inches and weighs 23–25 pounds.
2. His hard coat is about 6 inches long, straight, and parted down the middle of his back, with much hair on his face and ears. He should be brushed and combed every other day, or he will be a matted mess. Pet coats can be trimmed so they don't drag on the floor. He is blue, gray, fawn, cream, or black. His ears, muzzle, and tail are dark. A little white on his chest is allowed. A puppy (up to eighteen months old) coat may be a very different color from a mature coat.
3. His ears prick up or fold forward. His eyes are brown with a lively expression. His nose is black. His teeth meet in a scissors or level bite.
4. Disqualifications: His nose should not be flesh colored or brown.

Health Problems

He is a healthy breed, prone to no real problems.

Cautions when Buying

Don't choose the most independent or most scrappy Skye puppy, and don't choose a timid puppy.

Smooth Fox Terrier

Fine for Novice Owners ✓
Good with Older, Considerate Children ✓
Small in Size ✓
Smooth Coat ✓

	High	Med	Low
EXERCISE REQUIRED ✓		▰	
TRIMMING/CLIPPING REQUIRED ✓			▰
AMOUNT OF SHEDDING		▰	
ACTIVITY INDOORS	▰		
EASE OF TRAINING			▰
SOCIABILITY WITH STRANGERS		▰	

Temperament

He is one of the boldest, most energetic, most impulsive of all terriers—indeed of all breeds. Untiringly active and playful, he has a special passion for ball-chasing. He can adapt to any home so long as he is made a full participant in all activities and given some exercise. Don't ever let this breed off the leash except in a safe, enclosed area, for he is an explorer and a fast, agile, independent chaser who *will* take off. He is highly adaptable and makes an excellent traveling companion. Some are friendly with strangers, some are reserved. He is scrappy with other animals. This stubborn little dog has a keen, mischievous sense of humor, and needs early, firm, ongoing obedience training—*come* and *stay* are essential commands. He should not be spoiled, because he does not like the role of lapdog and needs someone who can control him and keep him from becoming pugnacious. Never hit a terrier, and don't play aggressive games like wrestling or tug-of-war. He can guard his food and toys from family members who have not established leadership over him. He is a barker and a digger.

History

He was developed in England as a plucky hunting terrier who bolted foxes from their dens. Until recently, he and his cousin, the Wire Fox, were classified as the same breed, even though they had different ancestors. Now they are correctly classified as separate breeds, and even though the Smooth is the older of the two, he's not nearly as popular. Some Fox Terrier enthusiasts feel that the Smooth is more aggressive and more energetic. Today he is a com-

panion, but he can still dispatch vermin. This is the dog on the RCA Victor record label listening to "His Master's Voice." AKC popularity: 75th of 130 breeds.

Physical Features

1. He stands 14½–15½ inches and weighs 15–19 pounds.
2. His hard coat needs only a quick brushing once a week. He is mostly white with black and/or brown patches. His head is often black with tan cheeks.
3. His ears fold forward. His eyes are dark with a fiery expression. His nose is black. His teeth meet in a scissors or level bite. His tail is docked.
4. Disqualifications: His nose should not be white, red, or spotted. His ears should not prick up completely or halfway. His teeth should not meet in a badly overshot or undershot bite.

Health Problems

He is a long-lived breed (fifteen years) but susceptible to deafness, glaucoma, cataracts, skin conditions, and heart disease. Check puppies' and parents' hearing by clapping and whistling.

Cautions when Buying

Don't choose the boldest, scrappiest, most energetic, or most independent Smooth Fox puppy.

Soft-Coated Wheaten Terrier 9

Fine for Novice Owners ✓
Good with Children ✓
Medium in Size
Long Coat ✓

	High	Med	Low
EXERCISE REQUIRED ✓		🦴	
TRIMMING/CLIPPING REQUIRED ✓		🦴	
AMOUNT OF SHEDDING ✓			🦴
ACTIVITY INDOORS ✓		🦴	
EASE OF TRAINING ✓		🦴	
SOCIABILITY WITH STRANGERS ✓	🦴		

Temperament

He is steadier, gentler, and more congenial than other terriers. His personality is happy, playful, and good-natured. He's a great herder and ratter for a farm, but so adaptable that he does fine in the city if given enough daily companionship and exercise and allowed to participate in games and activities. He's friendly with strangers, occasionally a bit timid, and usually good with other animals. However, males tend to be quarrelsome with other male dogs. Bright and only mildly stubborn, he responds quite well to obedience training. An occasional dominant Wheaten could guard his food and toys from family members who have not established leadership over him. He likes to dig, and some bark a bit.

History

He originated in Ireland as a small-game hunter, a herding dog, and an all-around farm dog. Today he is a companion, but he can still dispatch vermin. AKC popularity: 68th of 130 breeds.

Physical Features

1. Males stand 18–19 inches, females stand 17–18 inches. He weighs 30–40 pounds.
2. His coat is silky and wavy, with much facial hair, and should be brushed and combed every other day. Straggly hairs should be scissored every three months. He is born black and his coat changes color as he grows: up to one year it may be too dark, perhaps even with black-tipped hairs, and not yet wavy; up to two years it may be too light; but by two years it should be properly wavy and wheaten colored.
3. His ears hang down. His eyes are brown with an interested expression. His nose is black. His teeth meet in a scissors or level bite. His tail is docked.
4. Faults: His nose should not be any color other than solid black. His eyes should not be yellow.

Health Problems

He is susceptible to hip dysplasia, PRA, and flea allergies. Buy only from OFA- and CERF-registered parents.

Cautions when Buying

Don't choose the most active Wheaten puppy or a timid puppy.

Staffordshire Bull Terrier

For Experienced Owners Only
Good with Older, Considerate Children
Small in Size
Smooth Coat

	High	Med	Low
EXERCISE REQUIRED		🦴	
TRIMMING/CLIPPING REQUIRED			🦴
AMOUNT OF SHEDDING		🦴	
ACTIVITY INDOORS	🦴		
EASE OF TRAINING			🦴
SOCIABILITY WITH STRANGERS	🦴		

Temperament

This hardy breed is quiet and docile, yet very active. He loves to run and play, but he's so adaptable that he's fine in the city if vigorously exercised. He's friendly with strangers when his family is present, but this is a rugged little dog who should be accustomed to people at an early age. He's usually fine with other family pets, but can be aggressive with strange dogs, although most Staff Bulls will only attack other Staff Bulls (or similar breeds) and will consider other breeds as beneath them. Still, he should not be allowed off the leash except in a safe, enclosed area. Although stubborn, he will respond to consistent obedience training that is firm but fair. Never hit this breed, and don't play aggressive games like wrestling or tug-of-war. This is a strong, capable dog with excellent judgment, but he needs supervision and control from an owner who can match his intelligence.

History

Developed in England from crosses of the light-style Bulldog and the Old English Terrier, he was once used for dog fighting, but this breed is definitely *not* the Pit Bull Terrier that has been making headlines. (See the American Staffordshire Terrier profile for more information.) The Staffordshire Bull is a home guardian and companion. AKC popularity: 103rd of 130 breeds.

Physical Features

1. He stands 14–16 inches and weighs 25–45 pounds.
2. His hard coat needs only a quick brushing once a week.

He is black, blue, red, fawn, or brindle (brownish with black stripes and flecks), or any of these colors with white.
3. His ears prick up halfway. His eyes are brown. His nose is black. His teeth meet in a scissors bite.
4. Faults: His nose should not be flesh colored. His ears should not hang down or prick completely up.
5. Disqualifications: He should not be liver (brown) or black-and-tan.

Health Problems

He is susceptible to cataracts and serious tumors.

Cautions when Buying

BE CAREFUL. This is a strong little breed with a rugged background, and if you buy from a poor breeder you could end up with a dominant, aggressive Staffordshire Bull. Be sure you're buying an AKC Staffordshire Bull Terrier. Don't choose the boldest, most rambunctious, or most independent Staffie Bull puppy, and don't choose a timid puppy.

Welsh Terrier

Fine for Novice Owners ✓
Good with Older, Considerate Children ✓
Small in Size ✓
Wiry Coat ✓

	High	Med	Low
EXERCISE REQUIRED ✓		▰	
TRIMMING/CLIPPING REQUIRED ✓	▰		
AMOUNT OF SHEDDING ✓			▰
ACTIVITY INDOORS	▰		
EASE OF TRAINING			▰
SOCIABILITY WITH STRANGERS			▰

Temperament

He's steadier and less excitable than some terriers but still spirited and plucky. He's fine in the city if exercised, played with, and given companionship. Usually reserved with strangers, he is especially suited to a young athletic person or couple. Many are not as quarrelsome with other animals as are some other terriers, but some are scrappy, so don't let him off the leash except in a safe, enclosed area, for he can be a fast, agile chaser. He's independent but sensitive—he'll respond only to obedience training that does not include jerking around. Never hit any terrier, and don't play aggressive games like wrestling or tug-of-war. He can guard his food and toys from family members who have not established leadership over him. Some can be diggers. Some bark, but most are not as noisy as other terriers. Some have touch shyness—that is, they startle if touched unexpectedly.

History

He originated in Wales as a hunter of otter, fox, and badger. Today he is a companion, but he can still dispatch vermin. AKC popularity: 79th of 130 breeds.

Physical Features

1. He stands 14½–15½ inches and weighs 18–20 pounds.
2. His coat is short and wiry, with bushy eyebrows and a beard, and should be brushed and combed twice a week. He needs scissoring and coat shaping every three months. For pets, coat shaping means electric clipping; for show dogs, it means stripping (plucking) the dead hairs out one by one, because electric clipping softens the coat and fades the color. A puppy is born black, but the black recedes from his head, chest, and legs by four months, leaving a tan or reddish brown dog with a black neck and back.
3. His ears fold forward. His eyes are dark with a steady, confident expression. His nose is black. His teeth meet in a scissors or level bite. His tail is docked.

Health Problems

He is susceptible to skin conditions and some eye problems.

Cautions when Buying

Don't choose the boldest, scrappiest, or most independent Welsh puppy, and don't choose a timid puppy.

West Highland White Terrier

Fine for Novice Owners ✓
Good with Older, Considerate Children ✓
Small in Size ✓
Unusual Coat (See Physical Features 2. below) ✓

	High	Med	Low
EXERCISE REQUIRED ✓		🦴	
TRIMMING/CLIPPING REQUIRED ✓	🦴		
AMOUNT OF SHEDDING ✓			🦴
ACTIVITY INDOORS	🦴		
EASE OF TRAINING ✓		🦴	
SOCIABILITY WITH STRANGERS ✓	🦴	🦴	

Temperament

This hardy little dog is easier to handle than and not as stubborn as many terriers. He is assertive and plucky but seldom aggressive—he may chase the family cat for fun, but he's not usually serious about it. He is adaptable and usually friendly with strangers and makes an excellent traveling companion. Some males can be scrappy with other males. Occasionally he's a bit moody, but his cheerfulness usually returns quickly. Because he's such a confident little dog, he doesn't really enjoy pampering. He's fine in the city if exercised enough, given companionship, and played with. Although independent, he responds to obedience training much better than most other terriers. He can suddenly snap if teased or annoyed. He is a barker and a digger.

History

He was developed in the Scottish Highlands as a vermin hunter. Today he is a companion, the second most popular terrier after the Miniature Schnauzer, but he can still dispatch vermin. AKC popularity: 27th of 130 breeds.

Physical Features

1. He stands 10–11 inches and weighs 15–19 pounds.
2. His coat is about 2 inches long, straight, and harsh, and should be brushed and combed twice a week. He needs scissoring and coat shaping every three months. For pets, coat shaping means electric clipping; for show dogs, it means stripping (plucking) the dead hairs out one by one, because electric clipping softens the coat and fades the color. He is solid white.
3. His ears prick up. His eyes are dark with a keen expression. His nose is black. His teeth meet in a scissors or level bite.
4. Faults: His coat should not be any color other than white. His nose should not be any color other than black.

Health Problems

He is a long-lived breed (fifteen years) but susceptible to skin conditions, jawbone calcification (CMO), copper toxicosis (liver disease), Legg-Perthes (a musculoskeletal disease), and hernias.

Cautions when Buying

Don't choose the boldest, scrappiest, or most independent Westie puppy, and don't choose a timid puppy.

Wire Fox Terrier

Fine for Novice Owners ✓
Good with Older, Considerate Children ✓
Small in Size ✓
Wiry Coat ✓

	High	Med	Low
EXERCISE REQUIRED ✓		✖	
TRIMMING/CLIPPING REQUIRED ✓	✖		
AMOUNT OF SHEDDING ✓			✖
ACTIVITY INDOORS	✖		
EASE OF TRAINING			✖
SOCIABILITY WITH STRANGERS		✖	

Temperament

He is one of the boldest, most energetic, most impulsive of all terriers—indeed of all breeds. Untiringly active and playful, he has a special passion for ball chasing. He can adapt to any home so long as he is made a full participant in all activities and given some exercise. However, don't ever let this breed off the leash except in a safe, enclosed area, for he is an explorer and a fast, agile, independent chaser who *will* take off. He is highly adaptable and makes an excellent traveling companion. Some are friendly with strangers, some are reserved. He is scrappy with other animals. This stubborn little dog has a keen, mischievous sense of humor and needs early, firm, ongoing obedience training—*come* and *stay* are essential commands. He should not be spoiled because he does not like the role of lapdog and needs someone who can control him and keep him from becoming pugnacious. Never hit a terrier, and don't play aggressive games like wrestling or tug-of-war. He can guard his food and toys from family members who have not established leadership over him. He is a barker and a digger.

History

He was developed in England as a plucky hunting terrier who bolted foxes from their dens. Until recently, he and his cousin, the Smooth Fox, were classified as the same breed, even though they had different ancestors. Now they are correctly classified as separate breeds, and even though the Wire is the newer of the two, he's more popular. Today he is a companion, but he can still dispatch vermin. AKC popularity: 47th of 130 breeds.

Physical Features

1. He stands 14½–15½ inches and weighs 15–19 pounds.
2. His coat is short and wiry, with bushy eyebrows and a beard, and should be brushed and combed twice a week. He needs scissoring and coat shaping every three months. For pets, coat shaping means electric clipping; for show dogs, it means stripping (plucking) the dead hairs out one by one, because electric clipping softens the coat and fades the color. He is mostly white, with black and/or tan patches (often as a saddle). Usually he has a tan head, although a puppy is often born with a black head that becomes tan by one year.
3. His ears fold forward. His eyes are dark with a fiery expression. His nose is black. His teeth meet in a scissors or level bite. His tail is docked.
4. Disqualifications: His nose should not be white, red, or spotted. His ears should not prick up completely or halfway. His teeth should not meet in a badly overshot or undershot bite.

Health Problems

He is a long-lived breed (fifteen years) but susceptible to deafness, glaucoma, cataracts, skin conditions, and heart disease. Check puppies' and parents' hearing by clapping and whistling.

Cautions when Buying

Don't choose the boldest, scrappiest, most energetic, or most independent Wire Fox puppy.

The Toy Group

Affenpinscher · Rough Brussels Griffon · Smooth Brussels Griffon
Longhaired Chihuahua and Smooth Chihuahua · English Toy Spaniel
Italian Greyhound · Japanese Chin · Maltese · Toy Manchester Terrier
Miniature Pinscher · Papillon · Pekingese · Pomeranian
Toy Poodle · Pug · Shih Tzu · Silky Terrier · Yorkshire Terrier

Affenpinscher (AFF-en-pin-sher)

Fine for Novice Owners
Good with Older, Considerate Children
Little in Size
Wiry Coat

	High	Med	Low
EXERCISE REQUIRED			🦴
TRIMMING/CLIPPING REQUIRED		🦴	
AMOUNT OF SHEDDING			🦴
ACTIVITY INDOORS	🦴		
EASE OF TRAINING			🦴
SOCIABILITY WITH STRANGERS		🦴	

Temperament

He is somewhat terrierlike in personality: spirited, spunky, hardier than many toys. He does like some exercise. Some are friendly with strangers, some are reserved, most are fine with other family pets. But he is high-strung (he trembles a lot), and if he thinks he's being attacked by a stranger or another dog, he gets very excitable and scrappy. He has a mind of his own, and without a firm hand can be very demanding, throwing tantrums or sulking when he cannot have his own way. Spoiling is not recommended for this breed, especially since he is so bright and does respond well to calm, consistent obedience training. Treating him like an intelligent, hardy dog rather than a helpless baby does wonders for his self-confidence. He can be snappy when annoyed or frightened. He sometimes guards his toys and food. He can be noisy and hard to housebreak.

History

He originated in Europe, probably in Germany. His name means "monkey terrier" because of his bushy black face and large eyes. Today he is a companion. AKC popularity: 115th of 130 breeds.

Physical Features

1. He stands under 10¼ inches and weighs under 8 pounds.
2. His coat is short and wiry, with bushy eyebrows and a beard, and should be brushed and combed twice a week. Straggly hairs should be scissored and dead hairs stripped twice a year. He's usually black, but can also be black-and-tan, red, gray, or other colors.
3. His ears may be cropped or left alone. His eyes are black with a piercing expression. His nose is black. His teeth meet in a level or undershot bite. His tail is docked.
4. Faults: He should not have any white markings.

Health Problems

He is susceptible to slipped stifle (a joint disorder that may require surgery), fractures (don't let him jump off high furniture), ulcers on his protruding eyes, and respiratory difficulties because of his short face. Avoid hot, stuffy conditions—don't leave him in a closed car. He should be protected from extreme cold and damp.

Cautions when Buying

Don't choose the scrappiest or most independent Affenpinscher puppy, and don't choose a timid puppy.

Rough Brussels Griffon (Griff-ON)

Fine for Novice Owners
Good with Older, Considerate Children
Little in Size
Wiry Coat

	High	Med	Low
EXERCISE REQUIRED			■
TRIMMING/CLIPPING REQUIRED	■		
AMOUNT OF SHEDDING			■
ACTIVITY INDOORS	■		
EASE OF TRAINING		■	
SOCIABILITY WITH STRANGERS		■	

Temperament

Like the related Affenpinscher, he is somewhat terrierlike in personality: spirited, spunky, hardier than many toys. He does like some exercise, if possible. Some are friendly with strangers, some are reserved, most are fine with other family pets. But he is fearless and if he thinks he's being attacked by a stranger or another dog he displays great ferociousness that is part bluff, part genuine. He has a mind of his own and can be demanding, assertive, and defiant, so spoiling is not recommended. He needs firm but gentle obedience training, for he is both bright and sensitive. Training him to walk calmly on a leash will take time and patience. If you are tolerant of some of his eccentricities yet firm about the general rules of the household, he can be pleasant and fun. He can guard his food and toys. He is quick to retaliate against anything he perceives as harsh discipline or a threat. He can be noisy and hard to housebreak.

History

He originated in Brussels, the capital of Belgium, as a cross of ratting terriers, Affenpinschers, Pugs, and Ruby Spaniels. He comes in both smooth-coated and rough-coated varieties. He is a companion. AKC popularity: 94th of 130 breeds, but this combines both Smooths and Roughs.

Physical Features

1. He stands about 10 inches and weighs 8–12 pounds.
2. His coat is short and wiry and should be brushed and combed twice a week. He needs scissoring and coat shaping every three months. For pets, coat shaping means electric clipping; for show dogs, it means stripping (plucking) the dead hairs out one by one, because electric clipping softens the coat and fades the color. He can be reddish brown with a black face, black and reddish brown mixed, black with reddish brown markings, solid reddish brown, or solid black.
3. His ears may be cropped or left naturally half-pricked. His eyes and nose are black. His teeth meet in an undershot bite. His tail is docked.
4. Disqualifications: His nose should not be flesh colored or spotted. His tongue should not stick out of his mouth. He should not have any white spots. His teeth should not meet in an overshot bite.

Health Problems

He is susceptible to slipped stifle (a joint disorder that may require surgery), lacerations on his protruding eyes, and respiratory difficulties and heatstroke because of his short face. Avoid hot, stuffy conditions—don't leave him in a closed car.

Cautions when Buying

Don't choose the boldest, scrappiest, most independent Brussels puppy, and don't choose a timid puppy.

Smooth Brussels Griffon (Griff-ON)

Fine for Novice Owners
Good with Older, Considerate Children
Little in Size
Smooth Coat

	High	Med	Low
EXERCISE REQUIRED			✦
TRIMMING/CLIPPING REQUIRED			✦
AMOUNT OF SHEDDING		✦	
ACTIVITY INDOORS	✦		
EASE OF TRAINING		✦	
SOCIABILITY WITH STRANGERS		✦	

Temperament

Like the related Affenpinscher, he is somewhat terrierlike in personality: spirited, spunky, hardier than many toys. He does like some exercise, if possible. Some are friendly with strangers, some are reserved, and most are fine with other family pets. He is fearless, and if he thinks he's being attacked by a stranger or another dog he displays great ferociousness that is part bluff, part genuine. He has a mind of his own and can be demanding, assertive, and defiant, so spoiling is not recommended. He needs firm but gentle obedience training, for he is both bright and sensitive. Training him to walk calmly on a leash will take time and patience. If you are tolerant of some of his eccentricities yet firm about the general rules of the household, he can be pleasant and fun. He can guard his food and toys. He is quick to retaliate against anything he perceives as harsh discipline or a threat. He can be noisy and hard to housebreak.

History

He originated in Brussels, the capital of Belgium, as a cross of ratting terriers, Affenpinschers, Pugs, and Ruby Spaniels. He comes in both smooth-coated and rough-coated varieties. Today he is a companion. AKC popularity: 94th of 130 breeds, but this combines both Smooths and Roughs.

Physical Features

1. He stands about 10 inches and weighs 8–12 pounds.
2. His coat is smooth and needs only a quick brushing once a week. He can be reddish brown with a black face, black and reddish brown mixed, black with reddish brown markings, or solid black.
3. His ears may be cropped or left naturally half-pricked. His eyes and nose are black. His teeth meet in an undershot bite. His tail is docked.
4. Disqualifications: His nose should not be flesh colored or spotted. His tongue should not stick out of his mouth. He should not have any white spots. His teeth should not meet in an overshot bite. He should not be solid black.

Health Problems

He is susceptible to slipped stifle (a joint disorder that may require surgery), fractures (don't let him jump off high furniture), ulcers on his protruding eyes, and respiratory difficulties because of his short face. Avoid hot, stuffy conditions—don't leave him in a closed car. He should be protected from extreme cold and damp.

Cautions when Buying

Don't choose the boldest, scrappiest, most independent Brussels puppy, and don't choose a timid puppy.

Longhaired Chihuahua and Smooth Chihuahua (Chih-WAH-wah)

These are varieties of the same breed.

Fine for Novice Owners
Good with Older, Considerate Children
Little in Size
Feathered Medium-Length Coat or Smooth Coat

	High	Med	Low
EXERCISE REQUIRED			🦴
TRIMMING/CLIPPING REQUIRED			🦴
AMOUNT OF SHEDDING		🦴	
ACTIVITY INDOORS	🦴		
EASE OF TRAINING		🦴	🦴
SOCIABILITY WITH STRANGERS			🦴

Temperament

He's the ultimate apartment dog who doesn't like the outdoors, especially cold or damp weather. He is alert, curious, lively, and highly companionable, but also temperamental. Some are bold, some are timid and nervous. Reserved and suspicious with strangers, he is a one-person dog with a special affinity for the elderly. He's jealous of other pets but gets along well with other Chihuahuas. Unlike some toys, he's not too much of a tyrant when spoiled; still, it makes him insecure, which he expresses by snapping if he thinks someone or something is threatening him. He really doesn't need much obedience training, but treating him like an intelligent, hardy dog rather than a helpless baby does wonders for his self-confidence. He can snap if startled or teased, and he can guard his toys and food. He can be noisy and hard to housebreak.

History

He originated in Mexico as the *Techichi* breed, was crossed with small hairless dogs, and takes his name from the Mexican state of Chihuahua. In Montezuma's time he was favored by Aztec rulers as a religious good-luck charm. Today he is a companion. AKC popularity: 16th of 130 breeds.

Physical Features

1. He stands about 5 inches and weighs 2–6 pounds.
2. His coat is always soft. The Smooth needs only a quick brushing once a week. The Longhaired has longer hair on his ears, chest, stomach, legs, and tail, and should be brushed and combed twice a week. Colors include white, blonde, fawn, black-and-tan, patched, and others.
3. His ears prick up, although the tips may droop slightly. His eyes and nose are dark in dark-colored dogs, lighter in light-colored dogs. His teeth meet in a scissors or level bite.
4. Disqualifications: His tail should not be docked or bobtailed. His ears should not hang down nor be cropped. He should not weigh over 6 pounds. The coat of the Longhaired should not be thin or bare.

Health Problems

He is a long-lived breed (fifteen years) but susceptible to slipped stifle (a joint disorder that may require surgery), fractures (don't let him jump off furniture), jawbone disorders, rheumatism, eye problems, heart disease, and tooth and gum weaknesses. Toy breeds should not be fed canned food exclusively; they require dry food for crunching. He is born with a soft spot on the top of his head that does not always fully close—a blow here could kill him.

Cautions when Buying

THERE ARE MANY CHIHUAHUAS AROUND—BE VERY CAREFUL. If you buy this breed from a poor breeder, you could end up with a sickly, snappish, bad-tempered Chihuahua. Look for a breeder with good-sized (but *not* oversized), sturdy, outgoing dogs. Don't buy ultra-tiny (teacup) Chihuahuas—health and temperament are often sacrificed for size, and these dogs invariably have more health and behavior problems. Don't choose a scrappy, timid, or frail Chihuahua puppy.

English Toy Spaniel

Fine for Novice Owners
Good with Older, Considerate Children
Little in Size
Feathered Medium-Length Coat

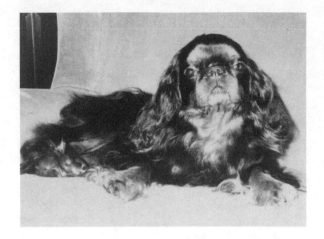

	High	Med	Low
EXERCISE REQUIRED			🦴
TRIMMING/CLIPPING REQUIRED			🦴
AMOUNT OF SHEDDING		🦴	
ACTIVITY INDOORS			🦴
EASE OF TRAINING		🦴	
SOCIABILITY WITH STRANGERS			🦴

Temperament

One of the most amiable and undemanding of the toys, this gentle little dog is very easy to live with. He is quiet and calm indoors, and although he does like to play, he doesn't like much outdoor exercise—a perfect apartment dog. He likes comfort and pampering and doesn't take nasty advantage of it. He's somewhat reserved with strangers but gets along well with other animals. Only mildly stubborn, he responds well to training but doesn't really need much. He's often a little timid—he cannot take rough handling or children's mischief. He wheezes a bit.

History

He probably originated in Asia, but he was developed in England and Scotland and was a favorite of the ruling class, including Mary Queen of Scots. He comes in four color varieties, each named after an English royal personage or noble house. You'll notice that one color variety is called the King Charles—don't confuse this color variety of English Toy Spaniel with the entirely different breed called the Cavalier King Charles Spaniel. The English Toy Spaniel was developed from hunting spaniels to be a companion. AKC popularity: 114th of 130 breeds.

Physical Features

1. He stands 9–10 inches and weighs 9–12 pounds.
2. His coat is silky, with longer hair on his ears, chest, stomach, legs, and tail, and should be brushed and combed twice a week. He comes in four color varieties:

the King Charles is black with rich tan markings above his eyes, on his muzzle, chest, and legs, and only a few white hairs on his chest are allowed; the Ruby is solid red with only a few white hairs on his chest allowed; the Blenheim is white with red patches and one dime-sized spot of red on his forehead; the Prince Charles is white with black and tan patches.
3. His ears hang down. His eyes are dark. His nose is upturned and black. His teeth meet in a level bite. His tail is docked to about 1½ inches.
4. Faults: His tongue should not stick out when his mouth is closed.
5. Disqualifications: A King Charles or Ruby variety should not have a big white patch on his chest, nor any white hairs on any other part of his body.

Health Problems

He is susceptible to slipped stifle (a joint disorder that may require surgery), lacerations on his protruding eyes, eyeball prolapse (the eye pops out of its socket because of a head blow), ear infections, and respiratory difficulties and heatstroke because of his pushed-in face. Avoid hot, stuffy conditions—don't leave him in a closed car. He should be protected from extreme cold and damp.

Cautions when Buying

Don't choose a timid English Toy Spaniel puppy.

Italian Greyhound

For Experienced Owners Only
Good with Older, Considerate Children
Small in Size
Smooth Coat

	High	Med	Low
EXERCISE REQUIRED			🦴
TRIMMING/CLIPPING REQUIRED			🦴
AMOUNT OF SHEDDING			🦴
ACTIVITY INDOORS		🦴	
EASE OF TRAINING		🦴	
SOCIABILITY WITH STRANGERS			🦴

Temperament

He is milder mannered, more peaceful, and much less demanding than are most toys. He's easy to live with and though he loves comfort, he's hardier than he looks. He likes a good run now and then in a safe, enclosed area; as might be expected from his sleek build, he is extremely fast and could be out of sight in minutes. He is well mannered and needs little obedience training, but training can work wonders for his self-confidence. Handle this sensitive, submissive little breed calmly and gently, but treat him like an intelligent, hardy dog rather than a helpless baby. He gets along well with other animals but is reserved and often timid with strangers. He could snap defensively if frightened, and he can be hard to housebreak.

History

An ancient breed found in the tombs of Egypt, he was popular with the Italians in the Middle Ages and was depicted in many Renaissance paintings. He was a favorite pet of European ruling families. He is a companion. AKC popularity: 70th of 130 breeds.

Physical Features

1. He stands 13–15 inches and weighs 7–13 pounds.
2. His coat is soft and needs only a quick brushing once a week. Colors include fawn, blue, black, red, and brown, often with white markings or patches.
3. His ears fold back against his head but rise at right angles to his head when he is excited. His eyes are dark.

His nose is black, dark gray, or brown. His teeth meet in a scissors bite.
4. Faults: His nose should not be light colored or two-toned. His ears should not prick up completely or fold forward. His tail should not curl into a ring over his back.
5. Disqualifications: He shoud not have brindle markings (brownish with black stripes and flecks). He should not have tan markings above his eyes, on his muzzle, and on his chest and feet (like a Doberman).

Health Problems

He is susceptible to slipped stifle (a joint disorder that may require surgery), fractures (don't let him jump off high furniture), PRA, and epilepsy. He should be protected from cold and damp.

Cautions when Buying

Don't choose a timid or frail Italian Greyhound puppy.

Japanese Chin

Fine for Novice Owners
Good with Older, Considerate Children
Little in Size
Feathered Medium-Length Coat

	High	Med	Low
EXERCISE REQUIRED			✦
TRIMMING/CLIPPING REQUIRED			✦
AMOUNT OF SHEDDING		✦	
ACTIVITY INDOORS		✦	
EASE OF TRAINING		✦	
SOCIABILITY WITH STRANGERS	✦		

Temperament

He is perky, proud, and playful—a perfect apartment dog who needs little exercise. He has an aristocratic demeanor, a mind of his own, and definite likes and dislikes. Bright, sensitive, and responsive, he needs little obedience training, but all dogs grow more self-confident with training and education, and this little breed responds very well to it. He likes comfort and could probably be pampered without becoming too much of a tyrant. He cannot take the roughhousing or mischief of young children, but he gets along well with strangers and other animals. He wheezes a bit.

History

He originated in China but was favored by Japanese nobility, who gave the little dogs as gifts to worthy foreigners. He was once called the Japanese Spaniel. There are several breeds of Oriental Chin-type dogs and there are several sizes of Japanese Chin, but only the smaller ones are considered correct for showing. He is a companion. AKC popularity: 73rd of 130 breeds.

Physical Features

1. He stands about 9 inches and weighs about 7 pounds.
2. His coat is silky with a thick ruff and profuse feathering on his thighs and tail; it should be brushed and combed twice a week. He's usually white with black patches, but occasionally his patches are red, lemon, orange, sable (brownish with black-tipped hairs), or brindle (brownish with black stripes and flecks.)
3. His ears hang down. His eyes are dark. His nose is black but may be reddish or flesh colored on dogs whose patches are not black. His teeth meet in a level or undershot bite. His tail is heavily feathered and curved over his back.
4. Disqualifications: If his patches are black, his nose should not be any color other than black.

Health Problems

He is susceptible to lacerations on his protruding eyes and respiratory difficulties and heatstroke because of his pushed-in face. Avoid hot, stuffy conditions—don't leave him in a closed car. He should be protected from extreme cold and damp.

Cautions when Buying

Don't choose a timid or scrappy Chin puppy.

Maltese (Mall-TESE)

Fine for Novice Owners
Good with Older, Considerate Children
Little in Size
Long Coat

	High	Med	Low
EXERCISE REQUIRED			✖
TRIMMING/CLIPPING REQUIRED			✖
AMOUNT OF SHEDDING			✖
ACTIVITY INDOORS	✖		
EASE OF TRAINING		✖	
SOCIABILITY WITH STRANGERS		✖	

Temperament

He is gentle, playful, and one of the brightest of the toys. He's spirited, likes to participate, and needs a lot of personal attention. He needs so little outdoor exercise that he makes a perfect apartment dog. He's reserved and sometimes timid with strangers but gets along well with other animals. He's better in an adult-only home because he wants to be the child in the family and doesn't like competition. Although his owners often pamper and spoil him, this can make him insecure, which he expresses by snapping at imagined threats. He's very responsive to nonforceful training and much hardier than he looks. Treating him like an intelligent, hardy dog rather than a helpless baby does wonders for his self-confidence. He can be noisy and hard to housebreak.

History

On the Mediterranean island of Malta he was a household pet of wealthy, cultured families and is probably a member of either the spaniel or the bichon family. He is a companion. AKC popularity: 25th of 130 breeds.

Physical Features

1. He stands about 8 inches and weighs 4–7 pounds.
2. His coat is straight and silky and needs to be brushed and combed every other day, or it will be a matted mess. Pet coats can be trimmed so they don't drag on the floor. He is solid white; fawn or lemon markings on his ears are allowed but usually disappear by the age of one year.
3. His ears hang down and are heavily feathered. His eyes are dark with a gentle expression. His nose is black. His teeth meet in a level or scissors bite. His tail is a long-haired plume carried over his back.

Health Problems

He's susceptible to slipped stifle (a joint disorder that may require surgery), eye infections, hypoglycemia (from not eating enough), and teeth and gum weaknesses. Toy breeds should not be fed canned food exclusively; they require dry food for crunching.

Cautions when Buying

BE CAREFUL. Many questionable breeders have exploited this breed. If you buy from a poor breeder you could end up with a sickly, snappish, noisy, hyperactive Maltese. Don't choose a scrappy, timid, or frail puppy. Look for a sturdy (but not oversized) puppy—the tinier (teacup) ones have more health and behavior problems.

Toy Manchester Terrier

Fine for Novice Owners
Good with Older, Considerate Children
Little in Size
Smooth Coat

	High	Med	Low
EXERCISE REQUIRED			▰
TRIMMING/CLIPPING REQUIRED			▰
AMOUNT OF SHEDDING			▰
ACTIVITY INDOORS	▰		
EASE OF TRAINING			▰
SOCIABILITY WITH STRANGERS			▰

Temperament

He's lively, gentle, and pleasant, but also high-strung, inquisitive, and demanding of attention. This is not a toy lapdog but a tiny terrier, and he does like some exercise and play. He is reserved and often timid with strangers and can be scrappy with other animals. Although he doesn't need much obedience training, he's made more confident by it. He's stubborn and sensitive but does respond fairly well to nonpunitive handling. He can snap when startled or irritated. He is a barker.

History

He was developed in Manchester, England, as a cross of Whippets and tough hunting terriers. There are two varieties of Manchester: the Standard, which is a member of the AKC Terrier Group, and the Toy, which is a member of the Toy Group. The Toy Manchester is a companion. AKC popularity: 84th of 130 breeds, but this includes both the Standard and Toy varieties.

Physical Features

1. He stands 8–13 inches and weighs up to 12 pounds.
2. His hard coat needs only a quick brushing once a week. He is black with rich tan markings above his eyes, on his cheeks, throat, legs, feet, and under his tail. There should be a black spot called a "thumb mark" on each front ankle, and narrow black stripes on all toes.
3. His ears prick up. His eyes are dark with a keen expres-sion. His nose is black. His teeth meet in a scissors or level bite.
4. Disqualifications: He should not be any color other than black-and-tan. He should not have a white stripe or patch measuring ½ inch. His ears should not be cropped.

Health Problems

He is susceptible to slipped stifle (a joint disorder that may require surgery), skin conditions, tooth and gum problems, bleeding disorders, and chills from cold and damp weather.

Cautions when Buying

Don't choose the most scrappy or independent Toy Manchester puppy, and don't choose a timid puppy.

Miniature Pinscher (PIN-sher)

For Experienced Owners Only
Good with Older, Considerate Children
Little in Size
Smooth Coat

	High	Med	Low
EXERCISE REQUIRED			▰
TRIMMING/CLIPPING REQUIRED			▰
AMOUNT OF SHEDDING			▰
ACTIVITY INDOORS	▰		
EASE OF TRAINING			▰
SOCIABILITY WITH STRANGERS			▰

Temperament

He's probably the most headstrong of the toys, and this is one of the reasons his fanciers like him so much. Spirited, animated, keen, confident, assertive, bold, proud—he thinks he's a big dog. He has a mind of his own and without a firm hand will be very demanding. He's a great apartment dog but does like some exercise. This is not a lapdog breed: He can be so much the temperamental tyrant when overly accommodated that spoiling is not recommended. He's reserved and sometimes suspicious with strangers and can be scrappy with other dogs. He can be high-strung and stubborn when it comes to obedience training, but he does need some if his owner is to keep the upper hand. If you treat him with respect but insist that he be as well mannered and responsive as a bigger dog, he will behave very well. He can snap when teased or annoyed, he can guard his toys and food, and he barks a lot.

History

He originated in Germany, but contrary to his appearance, he was not bred down from the Doberman but is actually a much older breed. The word *pinscher* simply means "terrier." He is a companion. AKC popularity: 39th of 130 breeds.

Physical Features

1. He stands 10–12½ inches and weighs 9–10 pounds.
2. His hard coat needs only a quick brushing once a week. He's usually red (actually brownish, sometimes inter-mingled with black hairs). Sometimes he's black with rust markings above his eyes, on his cheeks, lower jaw, throat, chest, legs, feet, and under his tail, and with narrow black stripes on his toes. Sometimes he's chocolate (dark brown) with the same rust markings.
3. His ears may be cropped or left alone. His eyes are dark. His nose is black on black and red dogs, brown on chocolate dogs. His teeth meet in a scissors bite. His tail is docked.
4. Disqualifications: He should not be any color other than red, black-and-tan, or chocolate-and-tan. He should not have white markings more than ½ inch wide. He should not have a patch of black hair surrounded by rust on either of his front ankles (this is called a "thumb mark" and is characteristic of the Toy Manchester Terrier). He should not stand over or under the height specification.

Health Problems

He is a healthy breed, prone to no real problems, but he should be protected from cold.

Cautions when Buying

Don't choose the boldest, most excitable, most energetic, or most independent MinPin puppy, and don't choose a timid puppy.

Papillon (PAP-ee-yon)

Fine for Novice Owners
Good with Older, Considerate Children
Little in Size
Feathered Medium-Length Coat

	High	Med	Low
EXERCISE REQUIRED			●
TRIMMING/CLIPPING REQUIRED		●	
AMOUNT OF SHEDDING		●	
ACTIVITY INDOORS		●	
EASE OF TRAINING	●		
SOCIABILITY WITH STRANGERS		●	

Temperament

He is one of the gentlest and most obedient of the toys. He needs little exercise and is a perfect apartment dog. When pampered and spoiled, he is dainty, nervous, and high-strung (but not usually a tyrant); when treated like a sturdy, intelligent, outgoing little dog, he's much more confident. He's bright and very responsive to obedience training. He gets along well with strangers and with other animals, although he can be a bit timid. He can snap if startled or annoyed, and he can be hard to housebreak.

History

He originated in France and was owned by such famous persons as Madame de Pompadour and Marie Antoinette. He was originally called the Dwarf Spaniel and had hanging ears, but when puppies began appearing with large, pricked, fringed ears that resembled the wings of a butterfly, the name became *Papillon*, French for "butterfly." He is a companion. AKC popularity: 55th of 130 breeds.

Physical Features

1. He stands 8–11 inches and weighs 8–10 pounds.
2. His coat is silky, with longer hair on his ears, chest, legs, and tail, and should be brushed and combed twice a week. Straggly hairs should be scissored every three months. He is white with patches of any color.
3. His ears should either hang down (called drop ears), or prick up (called erect ears). His eyes are dark with an alert expression. His nose is black. His teeth meet in a scissors bite. His tail is a long-haired plume arched over his back.
4. Faults: He should not have one ear up and one ear down. His ears should not be pricked up halfway. His ears should not be white, nor should the hair around his eyes be white. His nose should not be any color other than black.
5. Disqualifications: He should not stand over 12 inches. He should not be solid white, but he must have some white on him.

Health Problems

He is susceptible to slipped stifle (a joint disorder that may require surgery) and fractures (don't let him jump off high furniture).

Cautions when Buying

Don't choose a timid or frail Papillon puppy.

Pekingese (PEKE-in-ese)

Fine for Novice Owners
Good with Older, Considerate Children
Little in Size
Long Coat

	High	Med	Low
EXERCISE REQUIRED			🦴
TRIMMING/CLIPPING REQUIRED			🦴
AMOUNT OF SHEDDING		🦴	
ACTIVITY INDOORS			🦴
EASE OF TRAINING			🦴
SOCIABILITY WITH STRANGERS			🦴

Temperament

He is dignified, assertive, and independent, a big personality in a solid little body. Calm indoors and not needing much exercise, he's a perfect apartment dog. Some like being fussed over, while others are too proud to be anyone's lapdog. He's reserved with strangers and usually okay with other animals. He's willful and often demands his own way, but he's also sensitive and resents being jerked around or even scolded. This could make for almost impossible training, but fortunately he's fairly well mannered all on his own. Possessive of his food and toys, he could bite if irritated or pushed too far. Some Pekes are friendlier, more inquisitive, more outgoing, and more playful than others. He can be a barker. He wheezes.

History

He was considered sacred during the Chinese Tang Dynasty. When the Peking Imperial Palace was looted by the British in 1860, the breed was taken to England. He is a companion. AKC popularity: 17th of 130 breeds.

Physical Features

1. He stands 8–9 inches and weighs up to 14 pounds.
2. His coat is coarse with a thick ruff and profuse feathering on his ears, chest, legs, and tail and should be brushed and combed every other day or it will be a matted mess. He is red, fawn, black, white, black-and-tan, patched (called parti-color), sable (brownish with black-tipped hairs), or brindle (brownish with black

stripes and flecks). His face is always black, and there are spectacles around his eyes.
3. His ears hang down. His eyes are dark with a bold, independent expression. His nose is black. His teeth meet in a level bite. His tail is heavily feathered and curled over his back.
4. Disqualifications: He should not weigh over 14 pounds. His nose should not be flesh colored.

Health Problems

He is susceptible to lacerations and infections on his protruding eyes, eyeball prolapse (a head blow causes his eyes to pop out of their sockets), and eyelid/eyelash abnormalities. Respiratory difficulties and heatstroke are problems because of his pushed-in face, as are collapsed nostrils and cleft palate (both requiring sugery), jawbone disorders, spinal disc problems, and urinary stones. Avoid hot, stuffy conditions—don't leave him in a closed car. He should be protected from extreme cold and damp.

Cautions when Buying

Don't choose the most independent Peke puppy or a timid puppy.

Pomeranian

Fine for Novice Owners
Good with Older, Considerate Children
Little in Size
Thick Medium-Length Coat

	High	Med	Low
EXERCISE REQUIRED			🦴
TRIMMING/CLIPPING REQUIRED		🦴	
AMOUNT OF SHEDDING	🦴		
ACTIVITY INDOORS	🦴		
EASE OF TRAINING			🦴
SOCIABILITY WITH STRANGERS			🦴

Temperament

He is vivacious and spirited, bold and brash, one of the sturdiest and spunkiest of the toys. He likes outdoor exercise but can do nicely without it, so he is an ideal apartment dog. He is an inquisite busybody in the house, checking out every sound and activity. He's reserved with strangers and can be scrappy with other animals. He's very bright and responds well to obedience training if you're firm and sure of yourself, otherwise he'll do as he pleases. Although most enjoy pampering, he can become too demanding, so spoiling is not recommended for this independent little breed. He has a quick temper, can snap when irritated, can guard his food and toys, and likes to bark.

History

Descended from Icelandic sled dogs, he is a member of the spitz family. In Pomerania, Germany, he was bred down to his current size from thirty-pound herding dogs. Today he is a companion. AKC popularity: 14th of 130 breeds.

Physical Features

1. He stands 6–7 inches and weighs 3–7 pounds.
2. His coat is harsh and upstanding, with a profuse neck ruff and long hair on his chest, legs, and tail, and should be brushed twice a week regularly, daily when he's shedding. Straggly hairs should be scissored every three months. He's usually a solid color (red, white, black, brown, blue, orange, cream), sometimes with black shadings. Sometimes he is parti-color (white with color patches), sable (tannish gold with black-tipped hairs), or black with tan markings.
3. His ears prick up. His eyes are dark with a foxlike expression. His nose is usually black but is brown on brown dogs and blue on blue dogs. His teeth meet in a scissors bite. His tail is profusely feathered and carried over his back.
4. Faults: A solid-colored dog should not have a white chest, foot, or leg.

Health Problems

He is susceptible to slipped stifle (a joint disorder that may require surgery), eye infections, collapsed windpipe, heart disease, skin conditions, and tooth and gum disorders. Toy breeds shouldn't eat canned food exclusively; they require dry food for crunching.

Cautions when Buying

THERE ARE MANY POMERANIANS AROUND—BE VERY CAREFUL. If you buy this breed from a poor breeder you could end up with a sickly, noisy, snappish, bad-tempered Pomeranian who is often oversized to boot. Don't choose the boldest, most excitable, or most independent Pom puppy, a timid puppy, or a puppy whose parents are ultra-tiny or ultra-large.

Toy Poodle

Fine for Novice Owners
Good with Older, Considerate Children
Little in Size
Curly Coat

	High	Med	Low
EXERCISE REQUIRED			🦴
TRIMMING/CLIPPING REQUIRED	🦴		
AMOUNT OF SHEDDING			🦴
ACTIVITY INDOORS	🦴		
EASE OF TRAINING	🦴		
SOCIABILITY WITH STRANGERS		🦴	

Temperament

He's usually pert and lively, but some are high-strung and excitable. He likes some exercise, if possible. He's usually reserved (sometimes timid) with strangers, so he should be accustomed to people and noises at an early age. He's fine with other animals. One of the brightest and most sensitive of breeds, he is extremely responsive to gentle obedience training and can learn almost anything. Because he's so capable and willing to please, it's a shame to spoil him because he can turn stubborn and sulky. When treated like the hardy, intelligent dog he is, he is far more confident. He can snap when startled or irritated, and he is a barker.

History

The Standard Poodle, oldest and largest of the three Poodle varieties, originated in Germany or Russia as a duck retriever. His name comes from the German *pudelin*, meaning "to splash in the water," and his coat was clipped to streamline his swimming. France developed him into a show dog and circus dog, and many people refer to him, incorrectly, as a "French" Poodle. He was bred down to the Miniature and Toy (the smallest) varieties. AKC popularity: 2nd of 130 breeds, but this includes all sizes.

Physical Features

1. He stands up to 10 inches and weighs 5–7 pounds.
2. His coat needs scissoring and coat shaping every six weeks. The most common clip for pet Poodles is the sporting or kennel clip: 1 inch of body hair, a head topknot and tail pompom, face, feet, and tail shaved short. This clip is simple, looks nice, and should be brushed and combed twice a week; unfortunately, show dogs must wear ridiculously elaborate clips. He must be a solid color: black, white, gray, blue, silver, cream, red, apricot, chocolate (dark brown), or café-au-lait (pale brown).
3. His ears hang down. His eyes are usually dark but are lighter in apricot and brown dogs. His nose is usually black, but is brown on apricot and brown dogs. His teeth meet in a scissors bite. His tail is docked.
4. Disqualifications: He should not be parti-colored (two entirely different colors). He should not stand over the height specification. To enter the show ring, he must be clipped into one of three approved styles.

Health Problems

He is susceptible to PRA, cataracts, glaucoma, eye and ear infections, digestive problems, skin conditions, diabetes, and heart disease. Buy only from CERF-registered parents.

Cautions when Buying

THERE ARE MANY TOY POODLES AROUND—BE VERY CAREFUL. If you buy this breed from a poor breeder, you could wind up with a sickly, nervous, snappish Toy Poodle. Don't buy tiny "teacup" Poodles, because temperament and health are often sacrificed for size, and these dogs invariably have more health and behavior problems. Don't choose a timid, frail, or excitable Poodle puppy.

Pug

Fine for Novice Owners
Good with Children
Small in Size
Smooth Coat

	High	Med	Low
EXERCISE REQUIRED			✦✦
TRIMMING/CLIPPING REQUIRED			✦✦
AMOUNT OF SHEDDING		✦✦	
ACTIVITY INDOORS			✦✦
EASE OF TRAINING		✦✦	
SOCIABILITY WITH STRANGERS		✦✦	

Temperament

He is one of the sturdiest and most stable of the toys. Good-humored, amiable, and pleasant, he can be both dignified and playful. He likes some outdoor exercise, if possible. Usually he gets along well with strangers and other pets, but some are one-person dogs, and some like to be the child in the family and can be jealous of toddlers or other pets. Although somewhat stubborn, he seldom gets into real mischief, so be gentle and don't jerk him around during the little obedience training he needs. He's reliable enough that he can usually be spoiled without becoming a real tyrant. He wheezes, snorts, and grunts; he gulps in air, so he can be gassy.

History

He originated in China as the pet of Buddhist monks. It is recorded that he saved the life of William, Prince of Orange, by barking to announce the invasion of the Spaniards. It's also recorded that he carried secret messages from Josephine to her husband Napoleon. His name may come from his facial resemblance to marmoset monkeys, which were called pugs; "pug" is also an old term of affection. He is a companion. AKC popularity: 30th of 130 breeds.

Physical Features

1. He stands 10–11 inches and weighs 14–18 pounds.
2. His coat is soft and needs only a quick brushing once a week. He is apricot-fawn (with a black muzzle, black ears, and a black line called a "trace" running along his back) or solid black. Occasionally he is silver (with black muzzle, ears, and trace).
3. His ears hang down. His eyes are dark with a soft expression. His nose is black. His teeth meet in a level or undershot bite. His tail is curled tightly over his back, ideally in a double curl.

Health Problems

He is susceptible to eye lacerations on his protruding eyes, eyeball prolapse (the eyes pop out of their sockets if he receives a blow to the head), and eyelid/eyelash abnormalities. Also problems are respiratory difficulties and heatstroke because of his pushed-in face, collapsed nostrils (requiring surgery), and heart problems. Avoid hot, stuffy conditions—don't leave him in a closed car.

Cautions when Buying

Don't choose the most independent Pug puppy, or a timid or excitable puppy.

Shih Tzu (SHEET-sue)

Fine for Novice Owners
Good with Older, Considerate Children
Little in Size
Long Coat

	High	Med	Low
EXERCISE REQUIRED			■
TRIMMING/CLIPPING REQUIRED			■
AMOUNT OF SHEDDING		■	
ACTIVITY INDOORS		■	
EASE OF TRAINING		■	
SOCIABILITY WITH STRANGERS		■	

Temperament

He is proud, assertive, feisty, and playful. Some like outdoor exercise, some don't. He needs companionship, attention, and comfort, and he can probably be pampered without becoming too much of a tyrant. He gets along well with strangers and other animals. Although stubborn, he responds well to patient, fair obedience training that does not include a lot of jerking around. He can snap if irritated or pushed too far, especially in hot weather. He can be hard to housebreak, and some wheeze.

History

He was cherished by the Chinese Tang and Ming dynasty imperial families. During their Revolution, the Chinese killed many of their dogs rather than let them fall into the hands of the invaders but some survived and were taken to England and then to the rest of the world. His name means lion. He is a companion. AKC popularity: 9th of 130 breeds.

Physical Features

1. He stands 8–11 inches and weighs 9–18 pounds.
2. His coat is luxurious and flowing and needs to be brushed and combed every other day or it will be a matted mess. Pet coats may be trimmed so they don't drag on the floor. All colors are allowed, including blends.
3. His ears hang down and are heavily feathered. His eyes are dark but may be lighter in liver (brown) dogs. His

nose is usually black, but is liver on liver dogs. His teeth meet in a level or undershot bite. His tail is heavily plumed and curved over his back.
4. Faults: He should not have any pink in his nose.

Health Problems

He is susceptible to slipped stifle (a joint disorder that may require surgery), eye lacerations on his protruding eyes, respiratory difficulties and heatstroke because of his pushed-in face, eyelid and eyelash abnormalities, gum and tooth problems, and ear infections. Avoid hot, stuffy conditions—don't leave him in a closed car. He should be protected from extreme cold and damp.

Cautions when Buying

THERE ARE MANY SHIH TZUS AROUND—BE VERY CAREFUL. If you buy from a poor breeder, you could wind up with a sickly, nervous, hyperactive, or snappish Shih Tzu. Don't choose the most independent Shih Tzu puppy or a timid or excitable puppy.

Silky Terrier

Fine for Novice Owners
Good with Older, Considerate Children
Little in Size
Long Coat

	High	Med	Low
EXERCISE REQUIRED			✦
TRIMMING/CLIPPING REQUIRED		✦	
AMOUNT OF SHEDDING			✦
ACTIVITY INDOORS	✦		
EASE OF TRAINING		✦	
SOCIABILITY WITH STRANGERS		✦	

Temperament

Because of his terrier background, the Silky is one of the sturdiest and hardiest of the toys. He's keen, lively, curious, feisty, demanding, and mentally and physically quick. Some are friendly with strangers, some are reserved, some are timid. He can be scrappy with other dogs and should be watched around small pets like cats and rabbits, for he can be an excitable chaser. He likes some exercise and makes an excellent traveling companion. Although stubborn, he's also bright—he responds well to nonpunitive obedience training. When treated like the hardy, intelligent dog he is, he is far more stable and confident. Territorial and possessive of his toys and food, he may snap if irritated. He can be hard to housebreak and he likes to bark.

History

Originating in Sydney, Australia, and once called the Sydney Silky, he's a blend of Australian Terrier and Yorkshire Terrier. He was once a vermin hunter, but today he is a companion. AKC popularity: 49th of 130 breeds.

Physical Features

1. He stands 8–10 inches and weighs 8–10 pounds.
2. His coat is 5–6 inches long, silky, and parted down the middle of his back with a topknot on his head. His coat needs to be brushed and combed every other day or it will be a matted mess. Straggly hairs should be scissored every three months. He is dark blue or silver-blue with tan markings on his head, chest, legs, feet, and under his tail.
3. His ears prick up. His eyes are dark with a keen expression. His nose is black. His teeth meet in a scissors bite. His tail is docked.
4. Faults: His eyes should not be light colored.

Health Problems

He is a healthy breed, prone to no real problems.

Cautions when Buying

Don't choose the most independent, excitable, or scrappy Silky puppy, and don't choose a timid puppy.

Yorkshire Terrier

Fine for Novice Owners
Good with Older, Considerate Children
Little in Size
Long Coat

	High	Med	Low
EXERCISE REQUIRED			🦴
TRIMMING/CLIPPING REQUIRED			🦴
AMOUNT OF SHEDDING			🦴
ACTIVITY INDOORS	🦴		
EASE OF TRAINING		🦴	🦴
SOCIABILITY WITH STRANGERS		🦴	

Temperament

There are two schools of thought on the Yorkshire: (1) he is vigorous and hardy; (2) he is a delicate lapdog who loves pampering. The owner's view of him has much to do with how an individual dog turns out. For certain, he is assertive, lively, clever, demanding of attention, and must participate in all activities; he doesn't like to be left alone. He needs little outdoor exercise because he gets enough in the house. He gets along fairly well with strangers, although some are timid. Some can be scrappy with other animals. He's stubborn and darts around on the leash, but he does respond to nonpunitive training; don't jerk him too much. He can snap if startled or irritated, and he can guard his toys and food. He can be noisy and hard to housebreak. *When overprotected*, this bright little breed is notorious for behavior problems.

History

Originally called the Scotch Terrier, he was brought to Yorkshire, England, by Scotch weavers. He is a companion. AKC popularity: 12th of 130 breeds.

Physical Features

1. He stands 7–9 inches and weighs 3–7 pounds.
2. His coat is straight and silky with a profuse "headfall" (long hair hanging straight down from his ears and muzzle). His coat should be brushed and combed every other day or it will be a matted mess. Pet coats can be trimmed so they don't drag on the floor. He is dark steel blue with rich tan markings on his head, headfall, chest, and legs. Puppies are born black-and-tan and the proper color develops gradually.
3. His ears prick up. His eyes are dark with a sparkling expression. His nose is black. His teeth meet in a scissors or level bite. His tail is docked.

Health Problems

He is susceptible to slipped stifle (a joint disorder that may require surgery), eye infections, and gum and teeth weaknesses. Toy breeds should not eat canned food exclusively; they need dry food for crunching. He should be protected from cold.

Cautions when Buying

THERE ARE MANY YORKIES AROUND—BE VERY CAREFUL. If you buy this breed from a poor breeder, you could wind up with a sickly, snappy, yappy, or hyperactive Yorkie. Don't buy tiny "teacup" Yorkies because temperament and health are often sacrificed for size, and these dogs have more health and behavior problems. Don't choose a timid, excitable, or frail Yorkie puppy.

The Non-Sporting Group

Bichon Frise · Boston Terrier · Bulldog
Rough Chow Chow and Smooth Chow Chow · Dalmatian · Finnish Spitz
French Bulldog · Keeshond · Lhasa Apso · Miniature Poodle
Standard Poodle · Schipperke · Tibetan Spaniel · Tibetan Terrier

Bichon Frise (BEE-shon Free-ZAY)

Fine for Novice Owners ✓
Good with Children ✓
Small in Size ✓
Curly Coat ✓

	High	Med	Low
EXERCISE REQUIRED ✓			🦴
TRIMMING/CLIPPING REQUIRED ✓	🦴		
AMOUNT OF SHEDDING ✓			🦴
ACTIVITY INDOORS	🦴		
EASE OF TRAINING	🦴		
SOCIABILITY WITH STRANGERS ✓	🦴		

Temperament

He is gentle, cheerful, and easy to live with. Also lively, playful, and curious, he likes to participate in games and activities. He gets along well with strangers and other animals. He's a willing little dog, very responsive to gentle obedience training. He can be overly sensitive to a sudden touch (called "touch shyness"), and he generally doesn't love the roughhousing of mischievous toddlers, so young children should be taught to be fair with him. He can be hard to housebreak.

History

He originated on the Canary Island of Tenerife, and his name is a corruption of the Barbichon Water Spaniel, one of his ancestors. He was an item of trade among Spanish sailors, a great favorite of the Italian and French nobility, a common street dog, an organ grinder's dog, and a circus dog. Finally the French took control of the breed again and added *Frise* (French for a curly, loopy fabric) to his name. Today he is a companion. AKC popularity: 29th of 130 breeds.

Physical Features

1. He stands 9–12 inches and weighs 7–12 pounds.
2. His coat is at least 2 inches long, plush, and velvety. He looks like a white powderpuff when properly groomed, like an unkempt white Poodle if not. His coat should be brushed and combed every other day or it will be a matted mess. He needs scissoring and coat shaping every six weeks. He is white but may have a few shadings of buff, cream, or apricot. A puppy may have many such shadings, most of which disappear by six months of age.
3. His ears hang down; when pulled forward, they reach about halfway along his muzzle. His eyes are dark with an inquisitive expression. His nose is black. His teeth meet in a scissors bite. His plumed tail curves over his back.
4. Faults: His eyes should not be yellow, blue, or gray.

Health Problems

He is susceptible to skin conditions, ear infections, cataracts, and corneal/retinal disorders.

Cautions when Buying

Don't choose a timid Bichon puppy.

151

Boston Terrier

Fine for Novice Owners
Good with Children
Small in Size
Smooth Coat

	High	Med	Low
EXERCISE REQUIRED			✖
TRIMMING/CLIPPING REQUIRED			✖
AMOUNT OF SHEDDING		✖	
ACTIVITY INDOORS		✖	✖
EASE OF TRAINING		✖	
SOCIABILITY WITH STRANGERS		✖	

Temperament

He is playful and high-spirited—an altogether dapper little dog. He likes some exercise and especially loves to play games and chase balls. Extremely sensitive to his owner's moods, he has a special affinity for the elderly. Some are a bit nervous or reserved with strangers, others are more outgoing and friendly. Males can be scrappy with other male dogs. He's mildly stubborn but responds well to gentle obedience training. Some Bostons bark. He wheezes and gulps in air, so he can be gassy; feeding him high-quality (not supermarket) food will prevent gassiness.

History

One of the few breeds developed in America, he's a cross between the English Bulldog and an old English terrier. His founding fanciers were centered in Boston, but his name is misleading because his appearance and temperament are not at all terrierlike. He is a companion. AKC popularity: 26th of 130 breeds.

Physical Features

1. He stands about 12–14 inches and is divided into three weight classes: lightweight (under 15 pounds); middleweight (15–20 pounds); and the uncommon heavyweight (20–25 pounds).
2. He has a shiny coat, which needs only a quick brushing once a week. He is brindle (brownish with black stripes and flecks) or black. He has white markings on his muzzle and forehead, around his neck, and on his legs and feet.
3. His ears may prick up naturally (called bat ears) or be cropped to stand up. His eyes are dark with a soft expression. His nose is black. His teeth meet in a level or undershot bite. His short tail may hang down straight or twist itself into a spiral shape called a screw tail.
4. Faults: He should not be solid or mostly white.
5. Disqualifications: He should not be solid black, black-and-tan, liver (brown), or mouse colored (gray). His nose should not be flesh colored. His tail should not be docked.

Health Problems

He is a long-lived breed (fifteen years) but is susceptible to infections and lacerations on his protruding eyes, eyelid/eyelash abnormalities, and cataracts. Also problems are respiratory difficulties and heatstroke because of his pushed-in face, and tumors. Avoid hot, stuffy conditions—don't leave him in a closed car. He should be protected from extremes of temperature.

Cautions when Buying

BE CAREFUL. Many questionable breeders have exploited this breed, and if you buy from one of them you could end up with a sickly, noisy, hyperactive Boston. Don't choose the most scrappy or excitable Boston puppy, and don't choose a timid puppy.

Bulldog

Fine for Novice Owners
Good with Children
Medium in Size
Smooth Coat

	High	Med	Low
EXERCISE REQUIRED			🦴
TRIMMING/CLIPPING REQUIRED			🦴
AMOUNT OF SHEDDING		🦴	
ACTIVITY INDOORS			🦴
EASE OF TRAINING		🦴	
SOCIABILITY WITH STRANGERS		🦴	

Temperament

He is amiable, reliable, peaceful, and very sweet. Puppies are frisky, but adults are dignified and don't like much exercise, except for walks in cool weather. Some are friendly with strangers, some are reserved. He's not a barking watchdog, but if his family is threatened, this courageous breed can move much faster than you might think. Also, it would take a tremendous amount of serious teasing to provoke him, but once really aroused, he can be a force to reckon with. He is good with other pets but can be aggressive with strange dogs. Although mildly stubborn, he is surprisingly sensitive and responds well to early, patient, persistent obedience training that does not include hitting or jerking. This breed remembers what he learns. He is unusually possessive of his food—he should be fed by himself and no one (person or animal) should approach his dish while he's eating. He wheezes and slobbers a lot; he gulps in air, so he can be gassy.

History

He originated in England as a bull fighter and is often called the English Bulldog. The ferocity has been bred out of him, and in its place a squat build, bowed legs, and a grossly pushed-in face have been bred in. His fanciers say that this upturned face enables the dog to breathe while holding onto a bull's nose, but this is nonsense—today's Bulldog has difficulty breathing while asleep, let alone chasing a bull. He is a companion. AKC popularity: 37th of 130 breeds.

Physical Features

1. He stands 14–15 inches and weighs 40–55 pounds.
2. He has a glossy coat, which needs only a quick brushing once a week. He is brindle, white, red, fawn, or piebald (patched). A small white patch on his chest is allowed.
3. His ears are rose ears (the bottom part of the ear is up, but the top part flops over sideways). His eyes are dark with a dignified expression. His nose is black. His teeth meet in an undershot bite. His short tail may hang down straight or twist itself into a spiral shape called a screw tail.
4. Faults: He should not be solid black.
5. Disqualifications: His nose should not be brown.

Health Problems

He is a short-lived breed (ten years) and extremely susceptible to heatstroke and respiratory difficulties. Ideally, he should live in an air-conditioned home. Avoid hot, stuffy conditions—don't leave him in a closed car. Also problems are eyelid and eyelash abnormalities, skin conditions, congenital heart disease and heart attack, and cleft palate and collapsed nostrils (both requiring surgery).

Cautions when Buying

Don't choose the boldest or most independent Bulldog puppy.

Rough Chow Chow and Smooth Chow Chow

These are varieties of the same breed. The Rough Chow Chow is pictured.

For Experienced Owners Only
Good with Older, Considerate Children
Medium to Large in Size
Thick Medium-Length Coat

	High	Med	Low
EXERCISE REQUIRED		✦	
TRIMMING/CLIPPING REQUIRED			✦
AMOUNT OF SHEDDING	✦		
ACTIVITY INDOORS			✦
EASE OF TRAINING			✦
SOCIABILITY WITH STRANGERS			✦

Temperament

He is independent, introverted, remote—a serious, dignified dog. He's fine in the city if exercised enough, but this is a very reserved, protective breed who should be accustomed to people at an early age. He's often a one-person dog, and although he may be standoffish even with his owner, he's always polite, and some Chows are much more affectionate than others. He's a powerful dog, although his thick coat makes him seem bigger than he really is. He can be aggressive with other animals. He's stubborn and needs firm, experienced obedience training, but this proud breed cannot be harshly forced to obey. Persuasive consistency works much better. Never hit this dog. He can refuse commands from family members who have not established leadership over him. He can guard his food, and he may bite if annoyed or if touched suddenly, especially in hot weather. Ongoing exercise, socialization, and supervision are essential with a Chow.

History

He was developed in China as a hunter of wolf and bear. His name has an unusual origin: It was the custom of English sailors to simply write "chow chow" to describe various Oriental curios and bric-a-brac in the ship's hold, and since dogs were often included in the items of trade, they became known as Chow Chow dogs. He is a home guardian and companion. AKC popularity: 6th of 130 breeds (the Rough Chow is far more popular than the Smooth).

Physical Features

1. He stands 18–20 inches and weighs 50–70 pounds.
2. His coat is coarse with a dense undercoat. The Rough variety has a longer coat with a heavy neck ruff. He should be brushed twice a week, but brush daily when he's shedding or he will be a matted mess. He is usually red or black, but may be cream, cinnamon, or blue. Light shadings on his ruff, tail, and legs are allowed.
3. His ears prick up. His eyes are dark. His nose is black but may be blue or gray on blue Chows. His teeth meet in a scissors bite. His tongue is blue-black. His tail curls over his back.
4. Disqualifications: His nose should not be any color other than those listed above, such as pink or spotted. His tongue should not be red, pink, or red-pink spotted. Neither ear should be tipped over or hanging.

Health Problems

He is susceptible to hip dysplasia, skin conditions, and eyelid abnormalities. Buy only from OFA-registered parents. With his thick coat, he doesn't like the heat.

Cautions when Buying

THERE ARE SO MANY CHOWS AROUND THAT YOU SHOULD BE VERY CAREFUL WHEN BUYING ONE. If you buy from a poor breeder or raise the dog incorrectly, you could end up with an aggressive Chow. Don't choose the boldest or most independent Chow puppy, and don't choose a timid puppy.

Dalmatian

For Experienced Owners Only
Good with Children
Medium to Large in Size
Smooth Coat

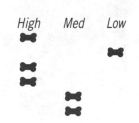

	High	Med	Low
EXERCISE REQUIRED	✦		
TRIMMING/CLIPPING REQUIRED			✦
AMOUNT OF SHEDDING	✦		
ACTIVITY INDOORS	✦		
EASE OF TRAINING		✦	
SOCIABILITY WITH STRANGERS		✦	

Temperament

He is a hardy, dependable gentleman but also high-spirited (sometimes high-strung) and playful (sometimes overexcitable), needing to participate in games and activities. He really belongs in the suburbs or country with an athletic owner who will give him vigorous daily exercise and regular romps. He can be restless and destructive if confined too much. He likes kids but may be too energetic for toddlers. Some are friendly with strangers, some are reserved, some are timid, some are protective; thus, he should be accustomed to people and noises at an early age. He is good with other pets and loves horses, but may be aggressive with strange dogs. He's so energetic that he should have early obedience training; although he's stubborn, under a firm, consistent hand he's capable of learning almost anything. Males can be roamers and explorers. This breed should not be kept outdoors in cold climates.

History

He takes his name from Dalmatia, a region around Austria. This was a versatile working dog: war sentinel, cart puller, sheepherder, vermin hunter, bird dog, circus performer, stable guardian, coach and carriage follower, and firehouse mascot. Today he is a companion. AKC popularity: 23rd of 130 breeds.

Physical Features

1. He stands 19–24 inches and weighs 46–65 pounds.
2. His hard coat needs brushing several times a week because he sheds year-round. He is white with black or liver (brown) spots, the spots sized between a dime and a half-dollar. A puppy is born all white and at two to six weeks develops his spots, which continue to grow in size until he's about six months old.
3. His ears hang down. His eyes are dark or blue in black-spotted dogs, light or blue in liver-spotted dogs. His nose is black on black-spotted dogs, brown on liver-spotted dogs. His teeth meet in a scissors bite.
4. Faults: His nose should not be spotted or flesh colored.
5. Disqualifications: His spots should not be any color other than black or liver, such as lemon yellow. He should not be tricolored (three colors). He should not have huge patches rather than spots (but a group of individual spots that cluster together are fine). He should not stand over the height specification. His teeth should not meet in an overshot or undershot bite.

Health Problems

He is susceptible to hip dysplasia, inherited deafness, skin conditions, and urinary stones. Buy only from OFA-registered parents. A low-protein diet is often recommended to avoid bladder and kidney stones in this breed.

Cautions when Buying

Don't choose the boldest, most independent, most excitable or energetic Dalmatian puppy, and don't choose a timid puppy.

Finnish Spitz

Fine for Novice Owners
Good with Children
Small to Medium in Size
Thick Medium-Length Coat

	High	Med	Low
EXERCISE REQUIRED		✦	
TRIMMING/CLIPPING REQUIRED			✦
AMOUNT OF SHEDDING	✦		
ACTIVITY INDOORS		✦	
EASE OF TRAINING		✦	
SOCIABILITY WITH STRANGERS			✦

Temperament

He is hardy, independent, and lively, but fine in the city if given enough daily exercise and if accustomed to people and noises at an early age. He's reserved with strangers, and some can be scrappy with other dogs. He likes to greet his family with the happy throaty sounds of crooning, purring, and yodeling. Strong-willed and stubborn, he's also sensitive to correction, so he needs firm but not harsh obedience training. He barks a great deal in a high-pitched voice. Because of his resemblance to the fox, he should be carefully supervised during hunting seasons in your area.

History

He is the national dog of Finland, his ancestors the hunting companions of ancient Finns. Spitz connotes a general type with husky-like features. Adapted from a small-game hunter into a bird hunter, he has an unusual hunting style: He flushes grouse into a tree, flicks his curled tail repeatedly to occupy the bird's attention, and barks in a high-pitched voice to alert his owner. He was called the Finnish Barking Bird-dog, and the Finns still hold an annual contest to crown their country's top barker. Americans don't appreciate barking bird dogs, so he was placed in the AKC Non-Sporting Group rather than the Sporting Group. He is a companion. AKC popularity: 113th of 130.

Physical Features

1. He stands 16–20 inches and weighs 24–35 pounds.
2. His coat is soft with a dense undercoat and needs brushing twice a week generally; daily when he's shedding. He is reddish gold, any shade from deep auburn to pale honey. White markings on his toes and chest, and a few black hairs on his tail and back are allowed. Puppies may have many black hairs, which disappear with age.
3. His ears prick up. His eyes are dark. His nose is black. His teeth meet in a scissors bite. His tail is bushy and curls over his back.
4. Faults: His nose should not be any color other than black. He should not have any white hairs, except on his toes and chest.

Health Problems

He is susceptible to hip dysplasia and PRA. Buy only from OFA- and CERF-registered parents.

Cautions when Buying

Don't choose the noisiest or most independent Finnish Spitz puppy or a timid puppy.

French Bulldog

Fine for Novice Owners
Good with Older, Considerate Children
Small in Size
Smooth Coat

	High	Med	Low
EXERCISE REQUIRED			🦴
TRIMMING/CLIPPING REQUIRED			🦴
AMOUNT OF SHEDDING		🦴	
ACTIVITY INDOORS		🦴	
EASE OF TRAINING		🦴	🦴
SOCIABILITY WITH STRANGERS		🦴	

Temperament

He is sweet, dependable, playful, outgoing, and easy to get along with. He's clean in the house and needs little outdoor exercise. He gets along fairly well with strangers and other animals, but he's often a one-person dog with a special affinity for the elderly. He needs a lot of companionship and cannot be owned and ignored. Although stubborn, he'll eventually respond to patient, consistent training that does not include hitting or harsh jerking. He can snap if annoyed, especially in hot weather. He makes quiet wheezing sounds and gulps in air, so he can be gassy. He doesn't bark very much. Be very careful with this breed around swimming pools—because of his squat build and heavy head, he cannot swim; many Frenchies are lost in drowning incidents.

History

He originated in England but was fully developed in France, a blend of toy English Bulldogs and other breeds. He is a companion and a good mouse hunter. AKC popularity: 87th of 130 breeds.

Physical Features

1. He stands about 12 inches and comes in two weight classes: the lightweight (about 19–22 pounds), and the heavyweight (22–28 pounds).
2. He has a smooth coat, which needs only a quick brushing once a week. Colors include brindle (brownish with black stripes and flecks), brindle-and-white, fawn, fawn-and-white, solid white, and a few others.
3. His ears prick up (bat ears). His eyes are dark with an alert expression. His nose is black, but may be lighter on light-colored dogs. His teeth meet in an undershot bite. His short tail may hang down straight or twist itself into a spiral shape called a screw tail.
4. Disqualifications: His ears should not be any type other than bat ears. He should not be solid black, black-and-white, black-and-tan, mouse colored (gray), or liver (brown). His eyes should not be two different colors. A dark-colored dog should not have a nose that is any color other than black. He should not have a harelip (a cleft in his lip). He should not weigh over 28 pounds.

Health Problems

He is susceptible to lacerations on his protruding eyes and respiratory difficulties and heatstroke because of his pushed-in face. Avoid hot, stuffy conditions—don't leave him in a closed car. He should be protected from extreme heat and cold.

Cautions when Buying

Don't choose the most independent Frenchie puppy or a timid puppy.

Keeshond (KAZE-hond)

Fine for Novice Owners
Good with Children
Medium in Size
Thick Medium-Length Coat

	High	Med	Low
EXERCISE REQUIRED		●	
TRIMMING/CLIPPING REQUIRED			●
AMOUNT OF SHEDDING	●		
ACTIVITY INDOORS		●	
EASE OF TRAINING		●	
SOCIABILITY WITH STRANGERS		●	

Temperament

He is gentler, quieter, and more sensible than some other spitz breeds. But he's still lively, needs companionship, and needs to participate. He's fine in the city if exercised daily. Most are friendly with strangers, but some are reserved, and some are timid, so he should be accustomed to people and noises at an early age. He gets along well with other animals. He can be somewhat high-strung and sensitive, but he responds fairly well to firm, patient obedience training that does not include a lot of jerking around. He has the unusual habit of baring his teeth in a happy smile, as a fox does. He can be hard to housebreak.

History

He originated in Holland as a riverboat watchdog and companion. When Holland divided itself into two political camps (the House of Orange and the House of the Patriots), the leader of the Patriots, Kees de Gysaeler, had a little dog also named Kees, who became the mascot of the Patriots. Eventually, though the Patriots lost the party battle, Kees's breed became the national dog of Holland. Today he is a companion. AKC popularity: 34th of 130 breeds.

Physical Features

1. He stands 17–18 inches and weighs 35–40 pounds.
2. His coat is harsh and upstanding with a profuse neck ruff and longer hair on his chest, legs, stomach, and tail. His coat needs brushing and combing twice a week regularly—daily when he's shedding or he will be a matted mess. He is a shaded mixture of gray and black with creamy legs, feet, and tail, and a black tip on the tail. Markings around his eyes form spectacles. A puppy is born black, turns cream by four months, and turns proper gray-black-cream by eighteen months.
3. His ears prick up. His eyes are dark brown. His nose is black. His teeth meet in a scissors bite. His tail is heavily coated and curls over his back.
4. Faults: His face should not be missing spectacles. His coat should not be parted down the middle of his back. His legs should not be black. His feet should not be white. He should not be any color other than gray/black with cream shadings; he should not be solid black or solid white, although some foreign registries allow solid colors.

Health Problems

He is susceptible to hip dysplasia, skin conditions and congenital heart disease. With his thick coat, he doesn't like the heat.

Cautions when Buying

Don't choose a timid Kees puppy.

Lhasa Apso (LAH-sa AHP-so)

For Experienced Owners Only
Good with Older, Considerate Children
Small in Size
Long Coat

	High	Med	Low
EXERCISE REQUIRED		▬	
TRIMMING/CLIPPING REQUIRED			▬
AMOUNT OF SHEDDING		▬	
ACTIVITY INDOORS	▬		
EASE OF TRAINING			▬
SOCIABILITY WITH STRANGERS			▬

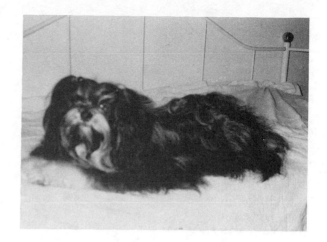

Temperament

He is keen, bold, assertive, independent, lively, and playful in a hardy, vigorous way. He's not a cuddly lapdog and should not be spoiled, for he can become a demanding tyrant. He is suspicious of strangers and should be accustomed to people at an early age. He can be jealous of other animals. Stubborn and strong-willed, he responds only to nonpunitive training—he must never be jerked or harshly disciplined. But note: Patient does not mean permissive, for this breed will take clever advantage of permissiveness. He's adaptable and makes an excellent traveling companion. He can refuse commands from family members who have not established leadership over him, and he can guard his toys and food. Don't play aggressive games like wrestling or tug-of-war, and never hit a Lhasa. He can have a quick temper and may snap if suddenly touched or pushed too far.

History

He originated in Tibet, a land of rugged terrain and climate. There he is called "Bark Lion Sentinel Dog." In village huts around the sacred city of Lhasa he barked when strangers approached, whereupon the huge Tibetan Mastiff would lumber over to investigate. The Buddhists believed that the Lord Buddha had tamed the lion to follow at his heels like a loyal dog; the lion thus has much symbolic significance and many Tibetan dogs are called "lion dogs." Today he is a companion. AKC popularity: 15th of 130 breeds.

Physical Features

1. He stands 9–11 inches and weighs 13–15 pounds.
2. His coat is straight and hard, his head and face overhung with hair. His coat needs to be brushed and combed every other day or he will be a matted mess. Pet coats can be trimmed so they don't drag on the floor. Colors include red, gold, cream, honey, white, gray, brown, black, parti-colored, and shaded.
3. His ears hang down and are heavily feathered. His eyes are dark brown. His nose is black. His teeth meet in a level or undershot bite. His tail is heavily feathered and carried over his back.

Health Problems

He is susceptible to kidney problems, skin conditions, and eye lacerations.

Cautions when Buying

THERE ARE MANY LHASAS AROUND—BE VERY CAREFUL. If you buy this breed from a poor breeder, you could end up with a sickly, snappish, bad-tempered, boss-dog Lhasa. Don't choose the boldest or most independent Lhasa puppy, and don't choose a timid, scrappy, or excitable puppy.

Miniature Poodle

9

Fine for Novice Owners ✓
Good with Older, Considerate Children ✓
Small in Size ✓
Curly Coat ✓

	High	Med	Low
EXERCISE REQUIRED ✓		🦴	
TRIMMING/CLIPPING REQUIRED ✓	🦴		🦴
AMOUNT OF SHEDDING ✓			
ACTIVITY INDOORS ✓	🦴	🦴	
EASE OF TRAINING ✓	🦴	🦴	
SOCIABILITY WITH STRANGERS		🦴	

Temperament

He is usually pert and lively, but some are high-strung and excitable. He likes some exercise. He's usually reserved (sometimes timid) with strangers, so he should be accustomed to people and noises at an early age. He's fine with other animals. One of the brightest and most sensitive of breeds, he is extremely responsive to gentle obedience training and can learn almost anything. Because he's so capable and willing to please, it's a shame to spoil him because he can turn stubborn and sulky. When treated like the hardy, intelligent dog he is, he is far more confident. He can snap when startled or irritated, and he is a barker.

History

The Standard Poodle, oldest and largest of the three Poodle varieties, originated in Germany or Russia as a duck retriever. His name comes from the German *pudelin*, meaning "to splash in the water," and his coat was clipped to streamline his swimming. France developed him into a show and circus dog, and many people refer to him, incorrectly, as a "French" Poodle. He was bred down into the Miniature and Toy varieties. AKC popularity: 2nd of 130 breeds, but this includes all three sizes.

Physical Features

1. He stands 10–15 inches and weighs 14–17 pounds.
2. The most common clip for pets is the sporting or kennel clip: 1 inch of body hair, a head topknot and tail pompom, face, feet, and tail shaved short. This clip,

which is simple and looks nice, needs brushing and combing twice a week; scissoring and expert coat shaping every six weeks. Unfortunately, show dogs must wear ridiculously elaborate clips. He must be a solid color: black, white, gray, blue, silver, cream, red, apricot, chocolate (dark brown), or café-au-lait (pale brown).

3. His ears hang down. His eyes are dark, but may be lighter in apricot and brown dogs. His nose is black, but may be brown on apricot and brown dogs. His teeth meet in a scissors bite. His tail is docked.
4. Disqualifications: He should not be parti-colored (two entirely different colors). He should not stand over or under the height specification. To enter the show ring, he must be clipped into one of three approved coat styles.

Health Problems

He is susceptible to PRA, cataracts, glaucoma, eye and ear infections, digestive problems, skin conditions, heart disease, and epilepsy. Buy only from CERF-registered parents.

Cautions when Buying

THERE ARE MANY MINIATURE POODLES AROUND—BE VERY CAREFUL. If you buy this breed from a poor breeder, you could end up with a sickly, nervous, snappy, yappy Poodle. Don't choose a timid or excitable Poodle puppy.

Standard Poodle

Fine for Novice Owners
Good with Children
Medium to Large in Size
Curly Coat

	High	Med	Low
EXERCISE REQUIRED		▸◂	
TRIMMING/CLIPPING REQUIRED	▸◂		
AMOUNT OF SHEDDING			▸◂
ACTIVITY INDOORS		▸◂	▸◂
EASE OF TRAINING	▸◂		
SOCIABILITY WITH STRANGERS		▸◂	

Temperament

He is usually lively, proud, sensible, and dignified, but some are high-strung or excitable. He's fine in the city if given enough daily exercise. He is usually reserved with strangers, but some are timid and some are protective, so he should be accustomed to people and noises at an early age. He's fine with other animals. One of the brightest and most sensitive of breeds, he is extremely responsive to obedience training and can learn almost anything. Because he is so capable and willing to please, it is a shame to spoil him because he can turn stubborn and sulky. This breed has almost no behavioral problems if dogs are treated like the hardy, intelligent animals they are.

History

The Standard Poodle, oldest and largest of the three Poodle varieties, originated in Germany or Russia as a duck retriever. His name comes from the German *pudelin,* meaning "to splash in the water," and his coat was clipped to streamline his swimming. France developed him into a show dog and circus dog, and many people refer to him, incorrectly, as a "French" Poodle. He is a companion. AKC popularity: 2nd of 130 breeds, but this statistic includes all three sizes.

Physical Features

1. He stands over 15 inches (usually 22–26 inches) and weighs 45–60 pounds.
2. The most common clip for pets is the sporting or kennel clip: 1 inch of body hair, a head topknot and tail pompom, face, feet, and tail shaved short. This clip, which is simple and looks nice, needs brushing and combing twice a week; scissoring and expert coat shaping every six weeks. Unfortunately, show dogs must wear ridiculously elaborate clips. He must be a solid color: black, white, gray, blue, silver, cream, red, apricot, chocolate (dark brown), or café-au-lait (pale brown).
3. His ears hang down. His eyes are dark, but may be lighter in apricot and brown dogs. His nose is black, but may be brown on apricot and brown dogs. His teeth meet in a scissors bite. His tail is docked.
4. Disqualifications: He should not be parti-colored (two entirely different colors). He should not stand under the height specification. To enter the show ring, he must be clipped into one of three approved coat styles.

Health Problems

He is susceptible to hip dysplasia, PRA, von Willebrand's Disease, bloat, cataracts, and sebaceous adenitis (severe skin disease). Buy only from OFA- and CERF-registered and VWD-tested parents.

Cautions when Buying

Most of the many Poodles around are Toys and Miniatures—the Standard has not been exploited so heavily by questionable breeders. Don't choose a timid or excitable Standard Poodle puppy.

Schipperke (SKIP-er-key)

Fine for Novice Owners
Good with Older, Considerate Children
Small in Size
Short Coat

	High	Med	Low
EXERCISE REQUIRED		🦴	
TRIMMING/CLIPPING REQUIRED			🦴
AMOUNT OF SHEDDING		🦴	
ACTIVITY INDOORS	🦴		
EASE OF TRAINING			🦴
SOCIABILITY WITH STRANGERS			🦴

Temperament

He is keen, bold, active, and hardy, a big-dog personality in a little-dog body. Curious and agile, he runs around checking the whole world out. He can adapt to any home so long as he is given enough exercise. He's usually good with other animals but suspicious with strangers; he should be accustomed to people at an early age. With children, some Schipperke fanciers say he's a superb companion for all ages, some say he's too quick-tempered to be completely reliable with toddlers. He is stubborn but eventually responds to firm, nonpunitive obedience training. He can guard his toys and food and snap if annoyed or startled. He can be hard to housebreak.

History

He originated in Belgium. He was a workmen's pet and barge watchdog, and his name is Flemish for "little captain." Today he is a companion and vermin hunter and gives a good effort as a home guardian. AKC popularity: 48th of 130 breeds.

Physical Features

1. He stands 12–13 inches and weighs up to 18 pounds.
2. His coat is short, harsh, and upstanding, with a dense neck ruff, and should be brushed once a week. He has very short hair on his face, ears, and legs. He is solid black.
3. His ears prick up. His eyes are dark brown with a sharp, questioning expression. His nose is black. His teeth

meet in a level or scissors bite. His tail is docked to no more than 1 inch long.
4. Faults: His eyes should not be light colored.
5. Disqualifications: He should not be any color other than solid black. His ears should not be hanging or only half-pricked. His teeth should not meet in a badly overshot or undershot bite.

Health Problems

He is a long-lived breed (fifteen years) and very healthy, prone to no real problems.

Cautions when Buying

Don't choose the boldest, most energetic, or most independent Schip puppy, and don't choose a timid or excitable puppy.

Tibetan Spaniel

Fine for Novice Owners
Good with Older, Considerate Children
Little in Size
Feathered Medium-Length Coat

	High	Med	Low
EXERCISE REQUIRED		●	
TRIMMING/CLIPPING REQUIRED			●
AMOUNT OF SHEDDING		●	
ACTIVITY INDOORS		●	
EASE OF TRAINING		●	●
SOCIABILITY WITH STRANGERS			●

Temperament

With his independent nature and quick agility, he has been compared to a cat. He is sturdy, happy, lively, and playful. He's reserved with strangers, usually good with other animals. Strong-willed and assertive but also sensitive, he reponds well to firm yet gentle obedience training. He likes to perch on high objects and watch over his family and home.

History

He is really not a spaniel at all, but he did originate in Tibet, a land of rugged terrain and climate. He sat atop monastery walls and barked when strangers approached the monastery, whereupon the huge Tibetan Mastiff would lumber over and investigate. The Buddhists believed that the Lord Buddha had tamed the lion to follow at his heels like a loyal dog; the lion thus has much symbolic significance and many Tibetan dogs are called "lion dogs." He is a companion. AKC popularity: 100th of 130 breeds.

Physical Features

1. He stands about 10 inches and weighs 9–15 pounds.
2. His coat is soft, with longer hair on his chest, stomach, legs, and tail, and should be brushed and combed twice a week. Colors include fawn, golden, cream, red, white, black, black-and-tan, patched, and shaded.
3. His ears hang down. His eyes are dark brown. His nose is black. His teeth meet in an undershot or level bite. His tail curls over his back.

4. Faults: His nose should not be brown or light colored. His coat should not be so long that there is no rectangle of daylight showing beneath him when he stands up.

Health Problems

He is a healthy breed, prone to no real problems, but because of his short face, beware of respiratory difficulties and heatstroke.

Cautions when Buying

Don't choose the most independent Tibetan puppy or a timid puppy.

Tibetan Terrier

Fine for Novice Owners
Good with Older, Considerate Children
Medium in Size
Long Coat

	High	Med	Low
EXERCISE REQUIRED		●	
TRIMMING/CLIPPING REQUIRED			●
AMOUNT OF SHEDDING			●
ACTIVITY INDOORS		●	
EASE OF TRAINING			●
SOCIABILITY WITH STRANGERS		●	

Temperament

He is lively, good-natured, and playful, but some are calmer than others. He adapts to any home so long as he is given daily exercise and companionship and is allowed to participate in games and activities. Some are friendly with strangers, some are reserved. He's usually good with other animals. This little dog can be stubborn, but he does respond to fair, consistent obedience training. He's adaptable and makes an excellent traveling companion. He can guard his food and toys from family members who have not established leadership over him. He can be hard to housebreak. He's usually quiet, but he has an unusual bark that starts deep, then rises like a siren.

History

He originated in Tibet, a land of rugged terrain and climate. He lived in the monasteries of the isolated Lost Valley, where the rare visitor was given one of these dogs for luck when he left for the arduous trip home. The Tibetan Terrier is not a terrier at all (he was given the name simply because of his size and shaggy coat), nor a guardian, nor a herder, nor a working dog. He is a companion. AKC popularity: 83rd of 130 breeds.

Physical Features

1. He stands 14–17 inches and weighs 20–30 pounds.
2. His coat may be straight or wavy, with much hair overhanging his head and face. He should be brushed and combed every other day or he will be a matted mess. Colors include white, gold, gray, silver, black, patched, and shaded.
3. His ears hang down and are heavily feathered. His eyes are dark. His nose is black. His teeth meet in a level or undershot bite. His tail is heavily feathered and curls over his back. His feet are webbed for a snowshoe effect.
4. Faults: His nose should not be any color other than black.

Health Problems

He is susceptible to hip dysplasia and PRA. Buy only from OFA- and CERF-registered parents.

Cautions when Buying

Don't choose the most active or most independent Tibetan puppy, and don't choose a timid puppy.

The Herding Group

Australian Cattle Dog · Bearded Collie · Belgian Malinois

Belgian Sheepdog · Belgian Tervuren · Bouvier des Flandres · Briard

Rough Collie · Smooth Collie · German Shepherd Dog

Old English Sheepdog · Puli · Shetland Sheepdog

Cardigan Welsh Corgi · Pembroke Welsh Corgi

Australian Cattle Dog

For Experienced Owners Only
Good with Older, Considerate Children
Medium in Size
Short Coat

	High	Med	Low
EXERCISE REQUIRED	▰		
TRIMMING/CLIPPING REQUIRED			▰
AMOUNT OF SHEDDING		▰	
ACTIVITY INDOORS		▰	
EASE OF TRAINING		▰	
SOCIABILITY WITH STRANGERS			▰

Temperament

He is a bold, hardy worker and a capable, watchful companion. He belongs in the suburbs or country where he can get the hard exercise he needs; he doesn't like to be cooped up with nothing to do. To keep his mind and body busy, he should work stock or be trained in obedience. Reserved and protective with strangers, he should be accustomed to people at an early age. He's smart but also clever at getting his own way; he's obedient but with an independent streak. He can be aggressive with other animals. He could bite if harshly treated, and he can nip at people's heels as though trying to herd them. Some can be barkers. Some are a bit timid or high-strung. Ongoing exercise, socialization, and activity are necessities with this breed.

History

He was developed in Australia to work wild cattle. Crosses between the Scottish Highland Collie, the wild Australian Dingo, the Dalmatian, and the Australian Kelpie produced a hardworking, protective dog often called the Queensland Blue Heeler, because he tends to bite at the heels of the cattle in order to drive and herd them. He is first a working stock dog, second a companion. AKC popularity: 64th of 130 breeds.

Physical Features

1. He stands 17–20 inches and weighs 35–50 pounds.
2. His coat is straight with a dense undercoat, and should be brushed once a week. He's usually blue-mottled (a mixture of blue, white, and gray), with tan markings on his head, chest, and legs, possibly with black markings on his head. He can also be red-speckled (white with many red flecks), possibly with darker red markings on his head. Puppies are born white.
3. His ears prick up. His eyes are dark brown with a suspicious expression. His nose is black. His teeth meet in a scissors bite.
4. Faults: A blue-mottled dog should not have any black markings on his body.

Health Problems

He is susceptible to hip dysplasia, PRA, and congenital deafness. Buy only from OFA-registered parents. Test puppies' hearing by clapping and whistling.

Cautions when Buying

If you're looking for an AuCaDo (AW-ca-doe) for a companion, don't buy from strict working lines, because this kind of dog may be too intense and businesslike to be at his best as a pet. Don't choose the boldest or most independent AuCaDo puppy, and don't choose a timid puppy.

Bearded Collie

Fine for Novice Owners
Good with Children
Medium in Size
Long Coat

	High	Med	Low
EXERCISE REQUIRED		●	
TRIMMING/CLIPPING REQUIRED			●
AMOUNT OF SHEDDING		●	
ACTIVITY INDOORS		●	
EASE OF TRAINING		●	
SOCIABILITY WITH STRANGERS		●	

Temperament

He is lively and playful, good-natured and easygoing, gentle and carefree. He's fine in the city if given plenty of daily exercise, but better in the suburbs or country where he can move. When not exercised enough or when not allowed to participate in games and activities, he may whine, dig, climb fences, and chew destructively. He is a sociable dog, mildly friendly with strangers, good with other animals. He has an independent streak but responds well to obedience training that is firm, fair, and not too repetitive, for he is easily bored and will show his sense of humor by disobeying in novel ways. He can nip lightly at people's heels as though trying to herd them.

History

He originated in Scotland as a sheepherder and cattle drover (driving cattle to market), and is sometimes called the Highland Collie. He is still a good herding dog as well as a companion. AKC popularity: 81st of 130 breeds.

Physical Features

1. Males stand 21–22 inches, females stand 20–21 inches. He weighs 45–55 pounds.
2. His coat is shaggy and parted down the middle of his back and should be brushed and combed every other day or he will be a matted mess. Puppies born black may stay black or lighten to gray or silver. Puppies born brown may darken to chocolate or lighten to sandy. Puppies born blue or fawn may also lighten. White markings are allowed on his head, chest, legs, feet, tail tip, and around his neck. Tan markings are allowed on his head and legs.
3. His ears hang down. His eyes and nose are dark or light, depending on coat color. His expression is soft. His teeth meet in a scissors bite.

Health Problems

He is susceptible to hip dysplasia. Buy only from OFA-registered parents.

Cautions when Buying

Don't choose the most energetic Beardie puppy or a timid puppy.

Belgian Malinois (MAL-in-wah)

For Experienced Owners Only
Good with Older, Considerate Children
Large in Size
Short Coat

	High	Med	Low
EXERCISE REQUIRED			
TRIMMING/CLIPPING REQUIRED			
AMOUNT OF SHEDDING			
ACTIVITY INDOORS			
EASE OF TRAINING			
SOCIABILITY WITH STRANGERS			

Temperament

He is alert, responsible, and serious. He doesn't like to sit still and tends to move in quick circles. He belongs in the suburbs or country with an athletic owner, for he can be restless without a lot of exercise and something to do. This watchful, territorial, protective breed should be accustomed to people at an early age. He is good with children if raised with children but is probably best with older kids. He can be aggressive with other dogs and should be watched around small pets like cats and rabbits. He respects and obeys a firm, experienced trainer, but he will not tolerate mistreatment. He can nip at people's heels as though trying to herd them. Some are more aggressive or high-strung than others—these could bite defensively or if provoked. He can refuse commands from family members who have not established leadership over him. This intelligent dog must be owned and handled by someone who is as smart and capable as he is.

History

He is one of three Belgian shepherd dogs who are so similar that most countries consider them varieties of the same breed. But the AKC divides them into three separate breeds. The Malinois is the least popular in America but the most popular in Belgium—in working trials he reigns supreme. He is used for herding, search and rescue, narcotics detection, Schutzhund, and as a companion. AKC popularity: 111th of 130 breeds.

Physical Features

1. Males stand 24–26 inches, females stand 22–24 inches. He weighs 60–75 pounds.
2. His coat is straight, with very short hair on his face, ears, and lower legs, and with a dense undercoat. His coat should be brushed once a week regularly—daily when he's shedding. He is rich fawn to mahogany brown, with black-tipped hairs. His face and ears are black. A little white is allowed on his chest and toes.
3. His ears prick up. His eyes are brown with an attentive expression. His nose is black. His teeth meet in a scissors or level bite.
4. Disqualifications: His ears should not hang down. His tail should not be stumpy or docked. He should not stand more than 1½ inches over or under the height specification.

Health Problems

He is susceptible to hip dysplasia. Buy only from OFA-registered parents.

Cautions when Buying

BE CAREFUL. If you buy this breed from a poor breeder or raise the dog incorrectly, you could end up with an aggressive Malinois. Avoid strict working lines if the dogs seem too intense or businesslike. Don't choose the boldest or most independent Malinois puppy, and don't choose a timid or suspicious puppy.

Belgian Sheepdog

For Experienced Owners Only
Good with Older, Considerate Children
Large in Size
Feathered Medium-Length Coat

	High	Med	Low
EXERCISE REQUIRED	🦴		
TRIMMING/CLIPPING REQUIRED			🦴
AMOUNT OF SHEDDING	🦴		
ACTIVITY INDOORS		🦴	
EASE OF TRAINING		🦴	
SOCIABILITY WITH STRANGERS			🦴

Temperament

He is alert, responsible, and serious. He doesn't like to sit still and tends to move in quick circles. He belongs in the suburbs or country with an athletic owner, for he can be restless without a lot of exercise and something to do. This is a watchful, territorial, protective breed who should be accustomed to people at an early age. He is good with children if raised with children, but is probably best with older kids. He can be aggressive with other dogs and should be watched around small pets like cats and rabbits. He respects and obeys a firm, experienced trainer, but he will not tolerate mistreatment. He can nip at people's heels as though trying to herd them. Some are more aggressive or high-strung than others—these could bite defensively or if provoked. He can refuse commands from family members who have not established leadership over him. This intelligent dog must be owned and handled by someone who is as smart and capable as he is.

History

He is one of three Belgian shepherd dogs who are so similar that most countries consider them varieties of the same breed. But the AKC divides them into three separate breeds. The Sheepdog (correctly called *Groenendael*, since all three are true sheepdogs) is seen more in the conformation ring than in the working or obedience ring, but he has been used for herding, search and rescue, narcotics detection, Schutzhund, and as a companion. AKC popularity: 82nd of 130 breeds.

Physical Features

1. Males stand 24–26 inches, females stand 22–24 inches. He weighs 60–75 pounds.
2. His coat is straight and abundant, with longer hair on his chest, stomach, legs, and tail, but very short hair on his face, ears, and lower legs. His coat should be brushed and combed twice a week regularly—daily when he's shedding. He is solid black with a little white allowed on his chest and the tips of his hind toes.
3. His ears prick up. His eyes are brown with an attentive expression. His nose is black. His teeth meet in a scissors or level bite.
4. Disqualifications: His ears should not hang down. His tail should not be stumpy or docked. He should not stand more than 1½ inches over or under the height specification.

Health Problems

He is susceptible to hip dysplasia. Buy only from OFA-registered parents.

Cautions when Buying

BE CAREFUL. If you buy this breed from a poor breeder or raise the dog incorrectly, you could end up with a shy, skittish, unstable Sheepdog. Don't choose the most independent Sheepdog puppy, and don't choose a timid puppy.

Belgian Tervuren (Ter-VER-en)

For Experienced Owners Only
Good with Older, Considerate Children
Large in Size
Feathered Medium-Length Coat

	High	Med	Low
EXERCISE REQUIRED	●		
TRIMMING/CLIPPING REQUIRED			●
AMOUNT OF SHEDDING	●		
ACTIVITY INDOORS		●	
EASE OF TRAINING		●	
SOCIABILITY WITH STRANGERS			●

Temperament

He is alert, responsible, and serious. He doesn't like to sit still and tends to move in quick circles. He belongs in the suburbs or country with an athletic owner, for he can be restless without a lot of exercise and something to do. This is a watchful, territorial, protective breed who should be accustomed to people at an early age. He is good with children if raised with children, but is probably best with older kids. He can be aggressive with other dogs and should be watched around small pets like cats and rabbits. He respects and obeys a firm, experienced trainer, but he will not tolerate mistreatment. He can nip at people's heels as though trying to herd them. Some are more aggressive or high-strung than others—these could bite defensively or if provoked. He can refuse commands from family members who have not established leadership over him. This intelligent dog must be owned and handled by someone who is as smart and capable as he is.

History

He is one of three Belgian shepherd dogs who are so similar that most countries consider them varieties of the same breed. But the AKC divides them into three separate breeds. The Tervuren excels at obedience and working trials and also does well in herding, search and rescue, narcotics detection, Schutzhund, and as a companion. AKC popularity: 91st of 130 breeds.

Physical Features

1. Males stand 24–26 inches, females stand 22–24 inches. He weighs 60–75 pounds.
2. His coat is straight, with longer hair on his chest, stomach, legs, and tail, but very short hair on his face, ears, and lower legs. His coat should be brushed and combed twice a week regularly—daily when he's shedding. He is rich fawn to mahogany brown with black-tipped hairs. His face is black. His ears are mostly black. A little white is allowed on his chest and toes.
3. His ears prick up. His eyes are brown with an attentive expression. His nose is black. His teeth meet in a scissors or level bite.
4. Disqualifications: His ears should not hang down. He should not stand more than 1 inch under or ½ inch over the height specification. He should not be solid black or solid brown. He should not have any white markings other than those listed above. His teeth should not meet in a badly undershot bite.

Health Problems

He is susceptible to hip dysplasia. Buy only from OFA-registered parents.

Cautions when Buying

BE CAREFUL. If you buy this breed from a poor breeder or raise the dog incorrectly, you could end up with a shy, skittish, unstable Tervuren. Don't choose the most independent Terv puppy, and don't choose a timid puppy.

Bouvier des Flandres (BOOV-yay day FLAHN-druh)

For Experienced Owners Only
Good with Children
Large to Giant in Size
Unusual Coat (See Physical Features 2. below)

	High	Med	Low
EXERCISE REQUIRED	✸		
TRIMMING/CLIPPING REQUIRED	✸		
AMOUNT OF SHEDDING			✸
ACTIVITY INDOORS			✸
EASE OF TRAINING		✸	
SOCIABILITY WITH STRANGERS			✸

Temperament

He is steady, responsible, and although rugged, usually docile. He does best in the suburbs or country where he can move and, preferably, do some type of work. Reserved and protective, he should be accustomed to people at an early age. He can be aggressive with strange dogs. This discerning, independent dog thinks for himself and trusts his own judgment, but he's not overly stubborn and responds very well to firm, fair obedience training. Many have an unusually keen sensitivity and scramble away from a sudden sound or movement, only to return immediately and fearlessly to investigate. He can nip at people's heels as though trying to herd them. He can refuse commands from family members who have not established leadership over him. This intelligent dog must be owned and handled by someone who is as smart and capable as he is.

History

He originated in Belgium, and his name comes from *voucheur de boeuf* (cattle driver) *des* (from) *Flandres* (Flanders). He was a farm dog who herded cattle and pulled milk carts. At the start of WWI, many Belgians were able to flee successfully when their Bouviers protected them against invading Germans. Some Belgians smuggled their dogs out to serve in the Belgian army. He is a herding dog, police and Schutzhund dog, home guardian, and companion. In Belgium he cannot be awarded a breed championship without a test of his working ability. AKC popularity: 56th of 130 breeds.

Physical Features

1. Males should stand 24½–27½ inches, females 23½–26½ inches. He weighs 65–100 pounds.
2. His coat is shaggy, harsh, and wiry, with bushy eyebrows and a beard and should be brushed and combed twice a week or he will be a matted mess. His coat needs scissoring and shaping every three months. For pets, coat shaping means electric clipping; for show dogs, it means stripping (plucking) the dead hairs out one by one, because electric clipping softens the coat and fades the color. He is usually black or gray, but may be fawn or brindle (brownish with black stripes and flecks). A little white on his chest is allowed.
3. His ears may be cropped or left alone. His eyes are dark brown with a bold expression. His nose is black. His teeth meet in a scissors bite. His tail is docked to the second or third joint.
4. Faults: He should not be brown, white, or particolored.

Health Problems

He is susceptible to hip dysplasia and bloat. Buy only from OFA-registered parents.

Cautions when Buying

BE CAREFUL. This is a big working breed, and if you buy from a poor breeder or raise the dog incorrectly, you could end up with a dominant, aggressive Bouvier. Don't choose the boldest or most independent Bouvier puppy or a timid puppy.

Briard (BREE-ard)

For Experienced Owners Only
Good with Children if Raised with Children
Large in Size
Long Coat

	High	Med	Low
EXERCISE REQUIRED	🦴		
TRIMMING/CLIPPING REQUIRED			🦴
AMOUNT OF SHEDDING		🦴	
ACTIVITY INDOORS		🦴	
EASE OF TRAINING			🦴
SOCIABILITY WITH STRANGERS			🦴

Temperament

He is spirited in body, independent and purposeful in mind. He belongs in the suburbs or country where he can move, for he needs much daily exercise and can be very restless if confined too much. Watchful and protective, he should be accustomed to people at an early age. Some are aggressive with other dogs. This is an energetic, responsible dog who knows his own mind and uses his own judgment; still, he remembers what he learns and he will respond to early, firm, consistent obedience training so long as he is not treated like a slave. He can nip at people's heels as though trying to herd them. He can refuse commands from family members who have not established leadership over him. This intelligent dog, like many of the herding breeds, must be owned and handled by someone who is as smart and capable as he is.

History

He originated in France as a livestock guardian, sheepherder, messenger dog, and military sentry. His name comes either from *Chien de Brie* (dog of Brie, a French province) or from *Chien d'Aubrey* (dog of Aubrey, a character in a popular French legend). Today he is still a herding dog and a companion. AKC popularity: 109th of 130 breeds.

Physical Features

1. Males stand 23–27 inches, females 22–25½ inches. He weighs 55–100 pounds.
2. His coat is at least 6 inches long and slightly wavy, with a dense undercoat, and should be brushed and combed every other day or he will be a matted mess. He is black, gray, tawny, or a combination of two of these colors. A few white hairs are allowed in his coat, and a small white spot is allowed on his chest.
3. His ears may be cropped or left hanging. His eyes are dark with a confident expression. His nose is black. His teeth meet in a scissors bite. He must have two dewclaws on each rear leg, and there may or may not be dewclaws on his front legs. His tail has a hook or crook at the end.
4. Disqualifications: His nose should not be any color other than black. His eyes should not be yellow or spotted. His tail should not be stumpy or docked. He should not be missing his two dewclaws on each rear leg. He should not be solid white or spotted. Any white spot on his chest should not be larger than 1 inch in diameter. He should not stand under the height specification.

Health Problems

He is susceptible to hip dysplasia, PRA, and bloat. Buy only from OFA- and CERF-registered parents.

Cautions when Buying

BE CAREFUL. This is a big, independent breed, and if you buy from a poor breeder or raise the dog incorrectly, you could end up with an aggressive, dominant Briard. Don't choose the boldest, most independent, or most energetic Briard puppy, and don't choose a timid puppy.

Rough Collie

Fine for Novice Owners
Good with Children if Raised with Children
Large in Size
Feathered Medium-Length Coat

	High	Med	Low
EXERCISE REQUIRED		✦	
TRIMMING/CLIPPING REQUIRED			✦
AMOUNT OF SHEDDING	✦		
ACTIVITY INDOORS			✦
EASE OF TRAINING		✦	
SOCIABILITY WITH STRANGERS		✦	

Temperament

He is usually proud, animated, and gentle, but some are nervous and high-strung. He's fine in the city if given enough daily exercise. Most are a bit cautious (sometimes timid) with strangers, so he should be accustomed to people and noises at an early age. He gets along well with other animals. Mildly stubborn and highly sensitive, he responds well to gentle, persuasive obedience training—don't discipline this sweet breed harshly or jerk him around. He can snap if startled or annoyed. He can nip at people's heels as though trying to herd them. Some are barkers. There is also a smooth-coated variety of Collie, far less popular than the Rough. Some Collie enthusiasts believe the Smooth has a steadier temperament and retains more working instincts.

History

He originated in Scotland as a black-and-white sheepherder. Queen Victoria helped develop the larger size, heavier coat, and various colors of today's Collie, and with his new, long, refined head, he looks very different from his working ancestor. Today he is a companion, but some can still herd. AKC popularity: 24th of 130 breeds, but this includes both Roughs and Smooths.

Physical Features

1. Males stand 24–26 inches, females stand 22–24 inches. He weighs 50–75 pounds.
2. His coat is straight and harsh, with longer hair on his chest, stomach, and legs, and should be brushed and combed every other day regularly, but daily when he's shedding or he will be a matted mess. Straggly hairs should be scissored every three months. He is sable (gold to mahogany, with white markings), tricolor (mostly black, with white markings and tan shadings), blue merle (a mottled blue-gray-black with white and/or tan markings), or white (actually mostly white, with sable, tricolor, or blue-merle markings).
3. His ears are three-quarters prick with the top quarter tipping forward. His eyes are dark, but one or both eyes may be blue or merle in blue-merle dogs; his expression is inquisitive. His nose is black. His teeth meet in a scissors bite.
4. Faults: His ears should not prick completely up.

Health Problems

He is susceptible to PRA, CEA (Collie Eye Anomaly), skin conditions, "collie nose" (a crusty nose caused by sensitivity to the sun), and congenital deafness (in blue merles and whites). Buy only from CERF-registered parents. Puppies' eyes must also have been individually examined by a certified ophthalmologist. Check the hearing of blue merles and whites by clapping and whistling.

Cautions when Buying

THERE ARE MANY ROUGH COLLIES AROUND—BE VERY CAREFUL. If you buy this breed from a poor breeder, you could end up with a high-strung, snappish, unreliable Collie. Don't choose a timid Collie puppy.

Smooth Collie

Fine for Novice Owners
Good with Children if Raised with Children
Large in Size
Short Coat

	High	Med	Low
EXERCISE REQUIRED		🦴	
TRIMMING/CLIPPING REQUIRED			🦴
AMOUNT OF SHEDDING	🦴		
ACTIVITY INDOORS			🦴
EASE OF TRAINING		🦴	
SOCIABILITY WITH STRANGERS		🦴	

Temperament

He is usually proud, animated, and gentle, but some are nervous and high-strung. He's fine in the city if given enough daily exercise. Most are a bit cautious (sometimes timid) with strangers, so he should be accustomed to people and noises at an early age. He gets along well with other animals. Mildly stubborn and highly sensitive, he responds well to gentle, persuasive obedience training—don't discipline this sweet breed harshly or jerk him around. He can snap if startled or annoyed. He can nip at people's heels as though trying to herd them. Some are barkers. There is also a rough-coated variety of Collie, which is much more popular than the Smooth. Some Collie enthusiasts believe the Smooth has a steadier temperament and retains more working instincts.

History

He originated in Scotland as a black-and-white sheepherder. Queen Victoria helped develop the larger size and various colors of today's Collie, and with his new, long, refined head, he looks very different from his working ancestor. Today he is a companion, but many can still herd. AKC popularity: 24th of 130 breeds (but this includes both Roughs and Smooths).

Physical Features

1. Males stand 24–26 inches, females stand 22–24 inches. He weighs 50–75 pounds.
2. His coat is short and harsh and should be brushed once a week regularly, daily when he's shedding. He is sable (gold to mahogany, with white markings), tricolor (mostly black, with white markings and tan shadings), blue merle (a mottled blue-gray-black, with white and/or tan markings), or white (actually mostly white, with sable, tricolor, or blue-merle markings).
3. His ears are three-quarters prick with the top quarter tipping forward. His eyes are dark, but one or both eyes may be blue or merle in blue-merle dogs; his expression is inquisitive. His nose is black. His teeth meet in a scissors bite.
4. Faults: His ears should not prick completely up.

Health Problems

He is susceptible to PRA, CEA (Collie Eye Anomaly), skin conditions, "collie nose" (a crusty nose caused by sensitivity to the sun), and congenital deafness (in blue merles and whites). Buy only from CERF-registered parents. Puppies' eyes must also have been individually examined by a certified ophthalmologist. Check the hearing of blue merles and whites by clapping and whistling.

Cautions when Buying

The Smooth Collie has not been exploited by poor breeders, so as long as you buy from a good breeder you should be fine. If you buy this breed from a poor breeder, you could end up with an extremely nervous, high-strung, snappish Collie. Don't choose a timid Collie puppy.

German Shepherd Dog

For Experienced Owners Only
Good with Children
Large in Size
Thick Medium-Length Coat

	High	Med	Low
EXERCISE REQUIRED		🦴	
TRIMMING/CLIPPING REQUIRED			🦴
AMOUNT OF SHEDDING	🦴		
ACTIVITY INDOORS		🦴	
EASE OF TRAINING	🦴	🦴	
SOCIABILITY WITH STRANGERS			🦴

Temperament

He is steady, discriminating, confident, very responsive to firm and fair obedience training, and willing and able to learn and do anything. Although fine in the city if given both physical and mental exercise (obedience training), some can be destructive when left alone too much. Reserved and protective, he should be accustomed to people at an early age. He can be aggressive with strange dogs. A dominant dog may refuse commands from family members who have not established leadership over him. This intelligent dog must be owned and handled by someone who is as smart and capable as he is. Anything less is a waste of this outstanding breed.

History

He was developed in Germany as a sheepherder, but he is now the world's leading guard, police, military, and Schutzhund dog. He is used for avalanche and earthquake search and rescue, narcotics and bomb detection, tracking missing persons, and guiding the blind. He's also a home guardian and companion. AKC popularity: 4th of 130 breeds.

Physical Features

1. Males stand 24–26 inches, females stand 22–24 inches. He weighs 65–100 pounds.
2. His coat is harsh with a dense undercoat and should be brushed several times a week because he sheds year-round. He is black-and-tan (tannish or reddish with a black back), bicolor (black with tan points, like a Doberman), golden sable (golden tan with black-tipped hairs), gray sable (steel or silver gray with black-tipped hairs), or solid black.
3. His ears prick up. His eyes are dark. His nose is black. His teeth meet in a scissors bite.
4. Faults: His coat should not be long. He should not be blue or liver (brown) in color.
5. Disqualifications: His ears should not be hanging. His teeth should not meet in an undershot bite. His nose should not be any color other than black. He should not be white.

Health Problems

He is susceptible to hip dysplasia, bloat, gastric disorders, panosteitis (a minor bone ailment of adolescents), spinal paralysis, pannus (a serious eye disease), and skin conditions. Buy only from OFA-registered parents.

Cautions when Buying

THERE ARE MANY SHEPHERDS AROUND—BE VERY CAREFUL. Show Shepherds are often spindly, long-bodied, and hyperactive. Working Shepherds are often from aggressive German lines; some are gigantic. Look for a breeder who strives for good temperament and health, some working ability in obedience competition, and sound conformation. If you buy this breed from a poor breeder or raise the dog incorrectly, you could end up with a sickly, nervous, or aggressive Shepherd. Don't choose the boldest or most independent Shepherd puppy, and don't choose a timid or hyperactive puppy.

Old English Sheepdog

For Experienced Owners Only
Good with Children
Large in Size
Long Coat

	High	Med	Low
EXERCISE REQUIRED	✖		
TRIMMING/CLIPPING REQUIRED			✖
AMOUNT OF SHEDDING		✖	
ACTIVITY INDOORS		✖	
EASE OF TRAINING			✖
SOCIABILITY WITH STRANGERS		✖	

Temperament

He is good-natured, sociable, and peaceful. He's enthusiastic, but usually not boisterous, although some are a bit bumptious or flighty. He does best in the suburbs or country, with daily exercise and space to move; he can be restless and flaky when confined too much. He's mildly friendly with strangers and usually good with other animals. He's stubborn but eventually responsive to firm, patient obedience training that does not include jerking around. This breed should never be hit or he could become nasty. He can nip at people's heels as though trying to herd them. Plenty of exercise and obedience training are essential to control this dog.

History

He originated in England, where he drove sheep and cattle to market. Since drover dogs were tax-exempt, their tails were docked as proof of their occupation. Today he is a companion. AKC popularity: 46th of 130 breeds.

Physical Features

1. He stands 21–25 inches and weighs 60–90 pounds.
2. His coat is shaggy with a dense undercoat and needs to be brushed and combed every other day or he will be a matted mess. Pet coats can be trimmed so they don't drag on the floor. He is gray, blue, grizzle (blue-gray), or blue merle (a mottled blue-gray-black), usually with white markings.
3. His ears hang down. His eyes are dark, but may be blue in blue dogs. His nose is black. His teeth meet in a level bite. His tail is a natural bobtail or docked under 2 inches.
4. Faults: He should not be brown or fawn.

Health Problems

He is susceptible to hip dysplasia, hereditary cataracts, auto-immune disorders, and skin conditions. Buy only from OFA- and CERF-registered parents.

Cautions when Buying

BE CAREFUL. This breed has been shamelessly exploited by questionable breeders, and if you buy from a poor breeder you could end up with a nervous, skittish, hyperactive, snappish, or aggressive Old English. Don't choose the most energetic, excitable, or independent Old English puppy, and don't choose a timid puppy.

Puli (POO-lee)

For Experienced Owners Only
Good with Older, Considerate Children
Small to Medium in Size
Long Coat

	High	Med	Low
EXERCISE REQUIRED		▰	
TRIMMING/CLIPPING REQUIRED			▰
AMOUNT OF SHEDDING			▰
ACTIVITY INDOORS	▰		
EASE OF TRAINING		▰	
SOCIABILITY WITH STRANGERS			▰

Temperament

He is keen, assertive, and curious. His quick, agile, dynamic movements have been likened to a bouncing spring. Some are nervous and high-strung, some are cheerful and outgoing. He's okay for the city if given enough daily exercise, but he likes the outdoors and can be restless if confined too much. This is a protective breed who should be accustomed to people at an early age. He is scrappy with other animals. He's extremely intelligent but in a clever, headstrong way—he needs firm, fair obedience training that does not include a lot of jerking and senseless repetition. Never hit this dog, for he has a quick temper and a quick bite. He may refuse commands from family members who have not established leadership over him. He can nip at people's heels as though trying to herd them. He is a barker. This intelligent dog must be owned and handled by someone who is as smart and capable as he is.

History

He originated in Hungary as a sheepherder and his name means "livestock driver." He is a tough, hardy worker, climbing across the backs of the flock to turn them in the right direction. Today he is a home guardian and companion. AKC popularity: 102nd of 130 breeds.

Physical Features

1. He stands 15–18 inches and weighs about 30 pounds.
2. The wiry hairs of his outercoat fuse with the woolly hairs of his undercoat to form real felt cords, which must occasionally be separated with the fingers; when the dog is shampooed, the cords must be rinsed well. If you choose to brush out the cords and trim the coat short, brush every other day, or he will be a matted mess. He may be shown either corded or brushed out. A puppy's coat falls into the proper cords as he matures. Usually he's dull black or rusty black, but he may also be solid gray or even solid white. A small white spot (less than 2 inches diameter) is allowed on his chest. His skin has a bluish or gray cast.
3. His ears hang down. His eyes are dark brown, with a keen expression. His nose is black. His teeth meet in a scissors bite. His tail curls over his back.

Health Problems

He is susceptible to hip dysplasia, and the hair should be pulled from his ears to avoid ear infections. Buy only from OFA-registered parents.

Cautions when Buying

BE CAREFUL. This is a quick, bright, independent breed, and if you buy from a poor breeder or raise the dog incorrectly, you could end up with a boss-dog Puli. Don't choose the boldest, scrappiest, most excitable, or most independent Puli puppy.

Shetland Sheepdog

Fine for Novice Owners
Good with Older, Considerate Children
Small in Size
Feathered Medium-Length Coat

	High	Med	Low
EXERCISE REQUIRED		●	
TRIMMING/CLIPPING REQUIRED		●	
AMOUNT OF SHEDDING	●		
ACTIVITY INDOORS		●	
EASE OF TRAINING	●		
SOCIABILITY WITH STRANGERS			●

Temperament

He is bright, gentle, highly sensitive, and a bit high-strung. He's lively, but fine in the city if given daily exercise. He is a swift runner and a graceful jumper. Reserved and often timid with strangers, he should be accustomed to people and noises at an early age. He gets along well with other animals. He learns quickly, is eager to please, and is very responsive to gentle obedience training. This sweet, willing little breed should never be manhandled. He can snap if startled, teased, or suddenly touched. He can nip at people's heels as though trying to herd them. He is a barker. Many breeders feel that the male is more cheerful and outgoing and makes a better family pet.

History

This small sheepherder originated on the rugged Shetland Islands off Scotland, where lack of food and space tended to produce miniature animals. He was not bred down in size from our present-day Collie, but from the same medium-sized, black-and-white working Collies that were the forerunners of our present-day Collie. Thus the smaller dog does not stem from the larger any more than the Shetland Pony stems from the Thoroughbred. Don't call him a "Toy" or "Miniature" Collie. He is a companion. AKC popularity: 13th of 130 breeds.

Physical Features

1. He stands 13–16 inches and weighs 14–18 pounds.
2. His coat is harsh with a dense undercoat and should be brushed and combed every other day regularly, daily when he's shedding or he will be a matted mess. Straggly hairs should be scissored every three months. He is sable (gold to mahogany with white markings), tricolor (black with white markings and tan shadings), or blue merle (mottled blue-gray-black with white and/or tan markings.)
3. His ears are three-quarters prick with the top quarter tipping forward. His eyes are dark, but one or both may be blue or merle in blue merle dogs. His nose is black. His teeth meet in a scissors bite.
4. Faults: His ears should not prick completely up. He should not be more than half white.
5. Disqualifications: He should not be brindle (brownish with black stripes and flecks). He should not stand over or under the height specification.

Health Problems

He is susceptible to PRA, CEA (Collie Eye Anomaly), heart disease, epilepsy, and deafness in blue merles. Buy only from CERF-registered parents. Puppies' eyes must have been individually examined by a certified ophthalmologist. Test the hearing of blue merles by clapping.

Cautions when Buying

THERE ARE MANY SHELTIES AROUND—BE VERY CAREFUL. If you buy this breed from a poor breeder, you could end up with a sickly, yappy, hyperactive Sheltie. Don't choose a timid Sheltie puppy.

Cardigan Welsh Corgi

Fine for Novice Owners
Good with Older, Considerate Children
Small in Size
Short Coat

	High	Med	Low
EXERCISE REQUIRED		●	
TRIMMING/CLIPPING REQUIRED			●
AMOUNT OF SHEDDING		●	
ACTIVITY INDOORS	●		
EASE OF TRAINING	●		
SOCIABILITY WITH STRANGERS		●	

Temperament

He is hardy, spirited, steady tempered, and easy to get along with. With a big-dog personality in a small-dog body, he can adapt to any home if given some exercise and allowed to participate in games and activities. Some are a bit one-personish, and most are a bit reserved with strangers. He is excellent with other Corgis, horses, livestock, and small pets, but he can be scrappy with strange dogs. He's very bright and responds well to obedience training. He's a good mouser and likes to work so much that, unless discouraged, he may nip at people's heels as if they were cattle. He makes an excellent traveling companion. Compared to the Pembroke Welsh Corgi, he is less excitable, more serious, and less accepting of strangers and other dogs. He can be a barker.

History

He originated in Cardiganshire, Wales, where he drove his master's cattle onto the common land to graze. He also drove off trespassing cattle and hunted vermin. He is related to the Dachshund, hence his low-slung build and slightly bowed front legs. Today he is a companion and makes a good effort as a home guardian. AKC popularity: 85th of 130 breeds.

Physical Features

1. He stands 10½–12½ inches and weighs 25–35 pounds.
2. His coat is harsh with a dense undercoat and should be brushed once a week. He is brindle (brownish with black stripes and flecks), black (sometimes with tan or brindle markings), blue merle (mottled blue-gray-black), sable (brownish with black-tipped hairs), or red. He usually has white markings.
3. His ears prick up. His eyes are dark, but one or both may be blue or merle in blue merle dogs; his expression is watchful. His nose is black, but may be spotted on blue merles. His teeth meet in a scissors bite.
4. Faults: His coat should not be long or fluffy.
5. Disqualifications: His nose should not be any color other than solid black (except on merles). His eyes should not be blue or partially blue (except in merles). His ears should not hang down. He should not be any color other than those listed. He should not be mostly white.

Health Problems

He is long-lived (fifteen years) but susceptible to PRA and glaucoma. Buy only from CERF-registered parents, and many breeders routinely X-ray for hip dysplasia as well.

Cautions when Buying

Don't choose the most independent Cardie puppy, and don't choose a timid or excitable puppy.

Pembroke Welsh Corgi

Fine for Novice Owners
Good with Older, Considerate Children
Small in Size
Short Coat

	High	Med	Low
EXERCISE REQUIRED		●	
TRIMMING/CLIPPING REQUIRED			●
AMOUNT OF SHEDDING		●	
ACTIVITY INDOORS	●		
EASE OF TRAINING	●		
SOCIABILITY WITH STRANGERS		●	

Temperament

He is hardy, spirited, steady-tempered, and easy to get along with. With a big-dog personality in a small-dog body, he can adapt to any home if given some exercise and allowed to participate in games and activities. Some are friendly with strangers, some are slightly reserved. He is excellent with other Corgis, horses, livestock, and small pets, and most are fine with strange dogs. He's very bright and responds well to obedience training. He's a good mouser and likes to work so much that, unless discouraged, he may nip at people's heels as if they were cattle. He makes an excellent traveling companion. Compared to the Cardigan Welsh Corgi, he is more excitable, more outgoing, and friendlier with strangers and other dogs. He can be a barker.

History

He originated in Pembrokeshire, Wales, where he drove his master's cattle out onto the common land to graze and drove off trespassing cattle. He is descended from the spitz family. Today he is a companion and makes a good effort as a home guardian. AKC popularity: 43rd of 130 breeds.

Physical Features

1. He stands 10–12 inches and weighs 25–30 pounds.
2. His coat is short with a dense undercoat and should be brushed once a week. He is red, sable (brown with black-tipped hairs), fawn, or black-and-tan. White markings are usual on his muzzle, face, around his neck, on his chest, and on his legs and feet.
3. His ears prick up. His eyes are brown with an interested expression. His nose is black. His teeth meet in a scissors bite. His tail is docked.
4. Faults: His eyes should not be black, yellow, or bluish. His ears should not be half-pricked or hanging. He should not be mostly white with a few markings of color. He should not have any white on his back or sides. He should not be black-and-white without some tan. He should not be bluish or grayish (usually with bluish gray eyes and nose). He should not have a long feathered coat (called a "fluffy").

Health Problems

He is long-lived (fifteen years) but susceptible to PRA, glaucoma, bleeding disorders, and back problems. Buy only from CERF-registered parents, and many breeders routinely X-ray for hip dysplasia as well.

Cautions when Buying

Don't choose a timid or excitable Pembroke puppy.

The Miscellaneous Group

Border Collie · Canaan Dog · Cavalier King Charles Spaniel

Hairless Chinese Crested and Powderpuff Chinese Crested

Chinese Shar-pei · Greater Swiss Mountain Dog · Miniature Bull Terrier

Border Collie

For Experienced Owners Only
Good with Older, Considerate Children
Medium in Size
Feathered Medium-Length Coat

	High	Med	Low
EXERCISE REQUIRED	▰		
TRIMMING/CLIPPING REQUIRED			▰
AMOUNT OF SHEDDING		▰	
ACTIVITY INDOORS	▰		
EASE OF TRAINING		▰	
SOCIABILITY WITH STRANGERS			▰

Temperament

He is a coiled spring of concentrated energy usually in motion when not under command. He's curious indoors, energetic and fast outdoors, and he belongs in the suburbs or country with an athletic owner who will work him on livestock or in obedience. When confined too much, he chews destructively and barks. He is a master escape artist and chaser of cars, bikes, joggers, cats, livestock, and deer. Usually reserved with strangers, some are timid and some are protective, so he should be accustomed to people at an early age. He can be aggressive with strange dogs. He develops unusual habits—one Border I know of stares for hours at the bathroom faucet, waiting for it to drip. He is capable of outstanding obedience work, but some are dominant and stubborn, and others are high-strung and sensitive, so training must be firm yet gentle. He can snap if teased or suddenly touched. This is not a breed for the casual owner—he requires ongoing exercise, supervision, activity, and training.

History

He may not be shown for championship points, only for exposure and for obedience titles, which he wins with a vengeance. He is also registered with the UKC. He herded sheep on the border between England and Scotland—he was named after the Colley Sheep and is also called Farm Collie and Working Collie. Pointer/Setter genes may have produced his famous "eye" (a fixed, hypnotic gaze as he crouches low and creeps up on the sheep). He excels in herding, narcotics and bomb detection, search and rescue, Schutzhund, and Frisbee competition. He is also a companion.

Physical Features

1. He stands 17–21 inches and weighs 35–50 pounds.
2. His coat is medium length with longer hair on his chest, stomach, legs, and tail. Some registries also allow a short coat. He should be brushed and combed twice a week. He's usually black-and-white, but may be gray-and-white, or blue-merle- (mottled blue-grey-black) and-white; sometimes he has tan shadings.
3. His ears usually prick up halfway. His eyes are dark, but some registries allow amber or blue. His teeth meet in a scissors or level bite. His front legs are shorter than his hind legs so that he seems to walk downhill.

Health Problems

He is susceptible to hip dysplasia, PRA, CEA (Collie Eye Anomaly), epilepsy, and congenital deafness. Buy only from OFA- and CERF-registered parents.

Cautions when Buying

BE CAREFUL. If you buy this breed from a poor breeder or raise the dog incorrectly, you could end up with an impossibly flaky Border Collie. Don't choose the boldest, most independent, or most energetic Border puppy, and don't choose a timid puppy.

Canaan Dog

Fine for Novice Owners
Good with Older, Considerate Children
Medium in Size
Short Coat

	High	Med	Low
EXERCISE REQUIRED		🦴	
TRIMMING/CLIPPING REQUIRED			🦴
AMOUNT OF SHEDDING		🦴	
ACTIVITY INDOORS		🦴	
EASE OF TRAINING	🦴		
SOCIABILITY WITH STRANGERS			🦴

Temperament

He is alert, lively, and watchful, with unusually keen senses and intuition. This is a docile, gentle dog. He can adapt to the city if given daily walks and occasionally taken someplace to romp. He is quite cautious (sometimes timid) with strangers, so he should be accustomed to people and noises at an early age. Some can be aggressive with strange dogs. Although he has an independent streak and is very sensitive to correction, he learns quickly and responds well to gentle obedience training that is not senselessly repetitive. Some bark and whine, and some are diggers.

History

He may not be shown for championship points, only for exposure and obedience titles. He is well known in Europe and his American fanciers are hoping for full AKC recognition in the future. He's the only breed native to Israel, where he was a watchdog, cattle guardian, sentry dog, and messenger dog. Today he is a companion, and some can herd.

Physical Features

1. He stands 19–24 inches and weighs 35–55 pounds.
2. His coat is about 1½ inches long and harsh and should be brushed once a week. Colors include predominantly white, with black, brown, or red patches, and a dark mask. He can also be solid tan, brown, reddish, liver, or black, with white trim. He can also be solid white.
3. His ears prick up. His eyes are dark but may be hazel in liver dogs. His nose is usually black but is brown on liver dogs. His teeth meet in a scissors bite. His tail curls over his back when he's excited.
4. Faults: His ears should not tip over or hang after one year of age. His coat should not be long.
5. Disqualifications: He should not be gray-striped or brindle (brownish with black stripes and flecks).

Health Problems

He is a healthy breed, susceptible to no real problems.

Cautions when Buying

Don't choose a timid Canaan puppy.

Cavalier King Charles Spaniel

Fine for Novice Owners
Good with Children
Small in Size
Feathered Medium-Length Coat

	High	Med	Low
EXERCISE REQUIRED		▰	
TRIMMING/CLIPPING REQUIRED			▰
AMOUNT OF SHEDDING		▰	
ACTIVITY INDOORS		▰	
EASE OF TRAINING	▰		
SOCIABILITY WITH STRANGERS	▰		

Temperament

He is sweet-tempered, gentle, outgoing, and playful. He has been called a sporting toy because he has a combination of spaniel and toy traits. He's fine in the city if given some exercise. He gets along well with strangers and other animals and needs to participate in activities; he doesn't like to be left alone for long periods of time. He's anxious to please and responds very well to gentle obedience training. As with any sweet-tempered breed, there may be a slight potential for timidity, so he should be accustomed to people and noises at an early age.

History

He may not be shown for championship points, only for exposure and obedience titles. He could have been fully recognized years ago, but his fanciers have always voted against full recognition because the AKC does not protect breed quality the way the strict National Cavalier Club does. Good breeders sell puppies with a Restricted-from-Breeding transfer: The puppy himself can be registered with the Club, but none of his offspring can. At twelve to eighteen months of age, if the dog is vet-certified to be free of the congenital health problems listed below, the restrictions are rescinded. The breed was developed in England, where he is among the five most popular breeds. He is a companion, and his national club participates in a worthy nursing home therapy program, using their little dogs to cheer up the elderly.

Physical Features

1. He stands 12–13 inches and weighs 13–18 pounds.
2. His coat is soft and silky, with longer hair on his ears, chest, stomach, legs, and tail, and should be brushed and combed every other day. He comes in four color varieties: the Blenheim is white with chestnut red markings, red ears, a white blaze, and, with luck, the much-prized red Blenheim spot on the top of the head; the Ruby is solid rich red; the Tricolor is white with black markings and rich red on the eyebrows, inside the ears, inside the legs, and under the tail; and the Black and Tan is mostly black, with chestnut markings in the same places as the Tricolor.
3. His ears hang down. His eyes are dark brown with a gentle expression. His nose is black. His teeth meet in a scissors, level, or slightly undershot bite.
4. Faults: His nose should not be flesh colored or spotted.

Health Problems

He is susceptible to heart disease, slipped stifle (a joint disorder that may require surgery), hip dysplasia, and eye problems. Buy only from OFA- and CERF-registered parents whose hearts have been checked.

Cautions when Buying

Don't choose a timid Cavalier puppy.

Hairless Chinese Crested and Powderpuff Chinese Crested

These are varieties of the same breed.

Fine for Novice Owners
Good with Older, Considerate Children
Little to Small in Size
Smooth Coat or Long Coat

	High	Med	Low
EXERCISE REQUIRED			🦴
TRIMMING/CLIPPING REQUIRED			🦴
AMOUNT OF SHEDDING			🦴
ACTIVITY INDOORS	🦴		
EASE OF TRAINING		🦴	
SOCIABILITY WITH STRANGERS	🦴		

Temperament

He is graceful, high-spirited, and happy. He climbs and jumps like a cat, and likes to grip his owner with his paws in a big hug. He makes an excellent apartment dog; although he is lively and needs exercise, he can get most of it in the house. He gets along well with strangers and other animals, but to avoid timidity, he should be accustomed to people and noises at an early age. Although independent, he is very bright and responds well to the gentle obedience training he needs to control his inquisitive activities. This little dog will be outgoing and confident if not treated like a fragile baby. Some are diggers.

History

He may not be shown for championship points, only for exposure and obedience titles. He was once fully recognized by the AKC, but interest waned to the point where he was dropped from recognition; he is now growing in popularity again and has been reinstated in Miscellaneous. He was discovered by Chinese traders either in Mexico or Africa. He belongs not to the domestic dog species of *Canis familiaris* but to *Canis africanis*. There is a lethal gene involved with the hairless mutation; when two of these genes come together, the dog is dead. Therefore, every surviving hairless dog carries one gene for hairless and one gene for puff. Thus, both hairless and puff varieties show up in the same litter, and are occasionally interbred to keep the crest on the hairless variety.

Physical Features

1. He stands 11 to 13 inches and weighs under 10 pounds.
2. The Hairless variety is hairless except for a tuft of hair on the head (the crest), the tail (plume), and the feet (sox). The Powderpuff has a long, straight, silky coat that should be brushed and combed every other day. He may be any color or combination, including black, white, brown, blue, pink, lilac, golden, and spotted. Sometimes his color seems to fade or intensify as the seasons change.
3. His ears prick up. His eyes and nose may be dark or light, depending on coat color. His teeth meet in a scissors or level bite.

Health Problems

The Hairless should be protected from extremes of weather but need not be babied (most will simply tan in the hot sun, but some will burn). He has only primitive dentition, so he must be fed dry food and hard biscuits to prevent his teeth from decaying. Some individuals have allergic skin reactions. Contrary to folklore, he does *not* have a higher body temperature than other breeds; he feels warm to the touch only because he does not have hair to provide insulation. The Powderpuff is prone to no real problems.

Cautions when Buying

Don't choose a timid Crested puppy.

Chinese Shar-pei

For Experienced Owners Only
Good with Older, Considerate Children
Medium in Size
Smooth Coat

	High	Med	Low
EXERCISE REQUIRED		●	
TRIMMING/CLIPPING REQUIRED			●
AMOUNT OF SHEDDING		●	
ACTIVITY INDOORS		●	
EASE OF TRAINING			●
SOCIABILITY WITH STRANGERS			

Temperament

He is dignified, serious, and quiet. Very clean indoors, he does well in the city if given daily walks and occasionally allowed to romp. However, this independent breed should not be let off the leash exept in a safe, enclosed area. Some are a bit standoffish even with their own families, but others are affectionate. Reserved and protective with strangers, he should be accustomed to people at an early age. He can be aggressive with other animals and must be carefully supervised in rural country, for he can run deer and molest livestock. He is stubborn and bold and needs firm obedience training to establish your leadership. He can refuse commands from family members who have not established leadership over him.

History

He may not be shown for championship points, only for exposure and obedience titles. He is a very old Chinese fighting dog; his loose skin enables him to twist and fight even when gripped by his opponent. Once nearly extinct, he is now quite popular and nearly ready for full AKC recognition. He is a companion.

Physical Features

1. He stands 18–20 inches and weighs 40–55 pounds.
2. His coat is always harsh and may be very smooth (called a "horse coat") or 1 inch long (called a "brush coat"). He needs only a quick brushing once a week. He must be a solid color such as black, cream, fawn, chocolate, or red. There may be darker shadings along his back and on his ears.
3. He may be a massive-headed, heavily wrinkled dog called a "meat mouth" or he may be a tighter-skinned type. His tiny ears fold forward. His eyes and nose are preferably dark, but may be lighter in light-colored dogs. His teeth meet in a scissors bite. His tongue, lips, and the roof of his mouth are blue-black, but may be lavender in light-colored dogs. His ringed tail is carried high.
4. Faults: His tongue should not be spotted. His coat should not be over 1 inch long.
5. Disqualifications: His ears should not prick up. His tongue should not be solid pink. He should not be white with pink eyes and nose. He should not be anything other than a solid color, for instance patched, brindle, or black-and-tan.

Health Problems

He is susceptible to hip dysplasia, eyelid abnormalities, and skin conditions. Buy only from OFA-registered parents.

Cautions when Buying

BE CAREFUL. This breed has been shamelessly exploited by questionable breeders, and if you buy from a poor breeder or raise the dog incorrectly, you could end up with a boss-dog Shar-pei. Don't choose the boldest or most independent Shar-pei puppy, and don't choose a timid puppy.

Greater Swiss Mountain Dog

Fine for Novice Owners
Good with Children
Giant in Size
Smooth Coat

	High	Med	Low
EXERCISE REQUIRED		🦴	
TRIMMING/CLIPPING REQUIRED			🦴
AMOUNT OF SHEDDING		🦴	
ACTIVITY INDOORS		🦴	
EASE OF TRAINING		🦴	
SOCIABILITY WITH STRANGERS	🦴		

Temperament

He is easygoing, patient, calm, and dependable. Although good indoors, he doesn't like confinement and belongs in the suburbs or country where he can have plenty of space and outdoor exercise. He should not be relegated to the backyard, for he needs daily companionship. He enjoys pulling carts or sleds. He gets along well with strangers and other animals, and he responds quite well to obedience training.

History

He may not be shown for championship points, only for exposure and obedience titles. He is descended from Roman mastiffs and is the largest and oldest of four Swiss "Sennenhunds" (dogs of the Alpine herdsmen). The smaller, longer-coated Bernese Mountain Dog is the only Sennenhund fully recognized by the AKC. The Greater Swiss, once called "Old Blaze," was a farmer's cart dog, pack dog, and home guardian who nearly vanished in Switzerland when the popular Saint Bernard moved in. Deservedly, he is now making a comeback. He is a companion.

Physical Features

1. He stands about 26–28 inches and weighs about 130 pounds.
2. His coat is short, about 1 inch or so, and should be brushed once a week. He is jet black with rust markings above his eyes and on his cheeks and legs; he must have a white blaze and white chest, usually he has white feet and tail tip, and sometimes he has a white patch on the back of his neck or a ring around his neck like a collar.
3. His ears hang down. His eyes are dark. His nose is black. His teeth meet in a scissors bite.
4. Faults: He should not be missing the white on his head or chest.

Health Problems

He is susceptible to hip dysplasia and bloat. Buy only from OFA-registered parents.

Cautions when Buying

Don't choose a timid Swissy puppy.

Miniature Bull Terrier

For Experienced Owners Only
Good with Older, Considerate Children
Small in Size
Smooth Coat

	High	Med	Low
EXERCISE REQUIRED		●	
TRIMMING/CLIPPING REQUIRED			●
AMOUNT OF SHEDDING		●	
ACTIVITY INDOORS	●		
EASE OF TRAINING			●
SOCIABILITY WITH STRANGERS		●	

Temperament

He is much like the Standard Bull Terrier: fiery, courageous, energetic. Some are bold and high-spirited, some are calmer and more laid-back. He's usually playful indoors and out. He needs daily exercise and companionship because he is easily bored and can be mischievous. He is a clever escape artist. Usually he's friendly with strangers, but some males are standoffish. The males are scrappy with strange dogs. He should be watched around small pets like cats and rabbits. Stubborn and independent, he needs firm, patient obedience training, but it helps if his trainer has a sense of humor because this dog has a quick-witted imagination and likes to invent new rules. He can refuse commands from family members who have not established leadership over him. He is a digger and may be hard to housebreak. This alert little breed must be owned by someone who can keep up with him—he is persistent in his need for time and attention.

History

He may not be shown for championship points, only for exposure and obedience titles. Bull Terriers have always come in several weight classes: Toy, Miniature, and Standard. Some time ago Toys and Miniatures were dropped as an allowed variety because of poor quality, but now Miniatures are back on the doorstep of full AKC recognition. He is a companion and an accomplished hunter of mice. See the Bull Terrier profile for more background.

Physical Features

1. He stands 10–14 inches and weighs 20–30 pounds.
2. His hard coat needs only a quick brushing once a week. He comes in two color varieties: *White* (either solid white or mostly white with a few color patches on his head) and *Colored* (preferably brindle, but often fawn, with white markings allowed on his head, chest, neck, feet, and tail tip).
3. His ears prick up. His eyes are dark with a keen expression. His nose is black—a pink nose on a white puppy darkens later. His teeth meet in a scissors or level bite.
4. Faults: A white dog should not have colored hairs anywhere except on his head. (Skin pigmentation is not considered colored hairs.)
5. Disqualifications: His eyes should not be blue. A colored dog should not be mostly white.

Health Problems

He is susceptible to deafness and lens luxation (eye disease). Check parents' and puppies' hearing by clapping.

Cautions when Buying

BE CAREFUL. This is a clever, independent breed, and if you buy from a poor breeder or raise the dog incorrectly, you could end up with a bossy Mini Bull. Don't choose the boldest, most boisterous, or most independent Mini Bull puppy or a timid puppy.

The Rare Breeds

American Eskimo Dog · American Hairless Terrier

Australian Shepherd · Bolognese · Cesky Terrier · Chinook

English Shepherd · Glen of Imaal Terrier · Havanese

Jack Russell Terrier · Kyi-Leo · Leonberger

Louisiana Catahoula Leopard Dog · Löwchen (Little Lion Dog)

Norwegian Buhund · Norwegian Lundehund (Puffin Dog)

Nova Scotia Duck Tolling Retriever · Polish Owczarek Nizinny Sheepdog

Shiba · Swedish Vallhund · Toy Fox Terrier

American Eskimo Dog

Fine for Novice Owners
Good with Older, Considerate Children
Little to Medium in Size
Thick Medium-Length Coat

	High	Med	Low
EXERCISE REQUIRED		🦴	
TRIMMING/CLIPPING REQUIRED		🦴	
AMOUNT OF SHEDDING	🦴		
ACTIVITY INDOORS	🦴		
EASE OF TRAINING	🦴	🦴	
SOCIABILITY WITH STRANGERS	🦴	🦴	

Temperament

He is keen, high-spirited, playful, and agile but fine in the city if given enough daily exercise and allowed to spend time outdoors. He likes companionship and he likes to participate in games and activities. Most are friendly with strangers and fine with other animals. Although very independent, he is bright and responds well to obedience training that is consistent but not harsh. Some can be barkers. Some are timid, so he should be accustomed to people and noises at an early age. Some display herding instincts. This is a happy, outgoing little dog who likes to have something to do.

History

He is recognized by the United Kennel Club (UKC) and American Eskimo Dogs of America. His ancestors include the German Spitz. This is a fairly popular little dog headed for AKC recognition. He is a companion.

Physical Features

1. He comes in three size varieties: Miniatures stand 12–15 inches and weigh 10–20 pounds. Standards stand 15–19 inches and weigh 25–35 pounds. Toys stand 9–12 inches and weigh 5–7 pounds.
2. His coat is harsh with a heavy neck ruff and longer hair on his chest, stomach, legs, and tail, with a dense undercoat. His coat should be brushed and combed twice a week regularly, but brush daily when he's shedding or he will be a matted mess. He is solid white, cream, or a mixture of white and biscuit cream.
3. His ears prick up. His eyes are black or brown with white eyelashes. A puppy's nose may have some pink, but must turn black or brown, although sometimes fading slightly in cold weather. His teeth meet in a scissors or level bite. His tail curls over his back.
4. Faults: His ears should not tip over. His nose, lips, and eyerims should not be pink. His tail should not have a double curl in it.
5. Disqualifications: He should not be any color other than white, biscuit, or cream. His eyes should not be blue.

Health Problems

He is a long-lived breed (fifteen years) and prone to no real problems.

Cautions when Buying

Don't choose the boldest, most energetic, most independent Eskie puppy, and don't choose a timid puppy.

American Hairless Terrier

Fine for Novice Owners
Good with Older, Considerate Children
Little in Size
Smooth Coat

	High	Med	Low
EXERCISE REQUIRED		●	
TRIMMING/CLIPPING REQUIRED			●
AMOUNT OF SHEDDING			●
ACTIVITY INDOORS	●		
EASE OF TRAINING		●	
SOCIABILITY WITH STRANGERS		●	

Temperament

The American Hairless is much more of a terrier than some other hairless breeds. He is lively, high-spirited, keen, and self-important. Some individuals are friendly, some are reserved, and some are timid, so he should be accustomed to people and noises at an early age. He gets along well with other dogs but should be watched around small pets like cats and rabbits. Although independent, he does respond fairly well to gentle, patient obedience training. Some are barkers and diggers.

History

He is not recognized by the AKC but by the American Hairless Terrier Association. This unique breed was created and is currently being developed by the Scott family of Louisiana. Hairless puppies that were popping up in their litters of wiry-coated Rat Terriers were bred together to create this completely hairless breed. He is a companion but can still do a good job on small vermin. This is a very new, very rare breed.

Physical Features

1. He comes in two size varieties: Toys stand 7½–10 inches and weigh 4–7 pounds; Miniatures stand 10–15 inches and weigh 7–14 pounds.
2. Unlike other hairless breeds, who do have some hair, he is completely hairless except for eyebrows and whiskers. His pink skin, warm to the touch, is mottled, spotted, or freckled with black, bronze, and/or red.

3. His ears prick up. His eyes and nose may be any color. His teeth meet in a scissors or level bite. His tail may be any natural length.
4. Faults: His ears should not tip over. He should not have any hair except whiskers and eyebrows. He should not be missing any teeth.

Health Problems

Unlike some other hairless breeds, he does not have any tooth or skin problems. But he is susceptible to sunburn and sunstroke.

Cautions when Buying

Don't choose a scrappy or timid American Hairless puppy.

Australian Shepherd

For Experienced Owners Only
Good with Children
Medium to Large in Size
Feathered Medium-Length Coat

	High	Med	Low
EXERCISE REQUIRED	●		
TRIMMING/CLIPPING REQUIRED		●	
AMOUNT OF SHEDDING		●	
ACTIVITY INDOORS		●	
EASE OF TRAINING	●	●	
SOCIABILITY WITH STRANGERS		●	

Physical Features

1. Males stand 20–23 inches, females stand 18–21 inches. He weighs 45–65 pounds.
2. His coat may be straight or wavy, with longer hair on his chest, neck, stomach, and legs, and should be brushed and combed twice a week. Straggly hairs should be scissored every three months. He is blue merle (mottled blue-gray-black), red (any shade from deep liver to light reddish brown), red merle (a patched or swirled or freckled red), or black. He has white and/or tan markings. A puppy's coat darkens as he matures.
3. His ears fold forward. His eyes may be brown, blue, amber, flecked, or mottled. His nose is black on black and blue merle dogs, brown on red and red merle dogs. A puppy's pink or spotted nose usually becomes solid by one year. His teeth meet in a scissors bite. His tail is a natural bobtail or docked to 4 inches.
4. Faults: His ears should not prick completely up.
5. Disqualifications: He should not be any color other than those listed. He should not have white body splashes. His nose should not be flesh colored. His bite should not be overshot or undershot.

Temperament

The working strain of Aussie is energetic, easily bored, and highly reflex-responsive (he'll scramble away if suddenly touched—this is a necessary survival instinct around kicking cattle, but a nuisance around the house). The pet strain is milder mannered and less sensitive. The Aussie needs plenty of exercise and work (either herding or obedience) and does best in the suburbs or country with an athletic owner. When confined and not kept busy he becomes restless and prone to destructive chewing. Some are friendly with strangers, some are reserved. He's usually good with other animals. Although he has an independent streak, he is highly intelligent and responds very well to obedience training. He may nip at people's heels as though trying to herd them. Plenty of exercise and activity are essential with an Aussie.

History

He is recognized by the United Kennel Club and also the Australian Shepherd Club of America (ASCA), which puts on its own shows and working trials. He comes from mixed English, Scottish, Spanish, and French Basque herding dogs, merged with Australia's blue bobtailed herders. He excels in herding, the Dogs for the Deaf program, narcotics detection, and search and rescue. He's also an agile frisbee competitor and a companion. This is a fairly popular breed, but most fanciers do not want AKC recognition for fear that his working instincts will be neglected in favor of appearance.

Health Problems

He is susceptible to hip dysplasia, PRA, cataracts, deafness in merles, and sunburn in reds and red merles. Buy only from OFA- and CERF-registered parents. Check the hearing of merles by clapping.

Cautions when Buying

If you're looking for a companion, avoid strict working lines, for these dogs may be too intense to be at their best as family pets.

Bolognese (Bole-oh-NESE)

Fine for Novice Owners
Good with Older, Considerate Children
Little in Size
Long Coat

	High	Med	Low
EXERCISE REQUIRED			🦴
TRIMMING/CLIPPING REQUIRED		🦴	
AMOUNT OF SHEDDING			🦴
ACTIVITY INDOORS	🦴		
EASE OF TRAINING		🦴	
SOCIABILITY WITH STRANGERS		🦴	

Temperament

He is vivacious, playful, and happy. Outdoors he's rough-and-tumble, indoors he shadows his owner possessively. He has been known to howl mournfully when his owner is busy and cannot pay attention to him. Most are friendly with strangers, but some are a bit timid, so he should be accustomed to people and noises at an early age. He gets along very well with other animals. This willing little dog is quite responsive to obedience training.

History

He is recognized by the Bolognese Club of America and the European FCI (Fédération Cynologique Internationale). He is closely related to the Bichon Frise and the Havanese. In the Italian city of Bologna he was the favorite of the Medici family, and was also owned by Catherine of Russia, the Queen of Austria, and the wealthy ladies-in-waiting of Belgian courts. He's shown widely in Europe and should become quite popular in the United States, but currently he's still a rare breed. He is a companion.

Physical Features

1. He stands 10–12 inches and weighs 10–12 pounds.
2. His coat is a mass of fluffy, flocky ringlets, either wavy or curly, and should be brushed and combed every other day or he will be a matted mess. Straggly hairs should be scissored every three months. He is white, sometimes with champagne shadings on his ears and back.
3. His ears hang down and are heavily feathered. His eyes are dark. His nose is black. His teeth meet in a scissors bite. His tail is plumed over his back.

Health Problems

He is a healthy breed, prone to no real problems.

Cautions when Buying

Don't choose a timid Bolognese puppy.

Cesky Terrier (CHES-key) 9

Fine for Novice Owners ✓
Good with Children ✓
Small in Size ✓
Unusual Coat (See Physical Features 2. below)

	High	Med	Low
EXERCISE REQUIRED ✓		🦴	
TRIMMING/CLIPPING REQUIRED ✓	🦴		
AMOUNT OF SHEDDING ✓			🦴
ACTIVITY INDOORS ✓		🦴	
EASE OF TRAINING ✓		🦴	
SOCIABILITY WITH STRANGERS ✓	🦴		

Temperament

He is a rugged little dog with more working instincts than some other terriers, but also with a more laid-back disposition. He is fairly calm indoors but playful and curious outdoors, and he does need some exercise. His personality is sweet, happy, and outgoing. He likes people and gets along well with other animals. This willing dog responds fairly well to obedience training. He is adaptable and makes an excellent traveling companion. He is quieter than most terriers, but some can bark a bit.

History

He is recognized by the Cesky Terrier Club of America and the European FCI. Also called the Czech or Bohemian Terrier, he was created by a Czechoslovakian geneticist and hunter who felt that the large heads, broad chests, and harsh coats of many working terriers made it difficult for them to slither underground after foxes and rats. He crossed the Scottish Terrier with the Sealyham Terrier to create the Cesky. The breed is popular in parts of northern Europe, but still rare in the United States. He is a companion and an avid hunter of small game and vermin.

Physical Features

1. He stands 11–14 inches and weighs 13–20 pounds.
2. His coat is lustrous, silky, and wavy, clipped short on his neck, back, and sides but left long on his head, chest, stomach, and legs. His coat needs brushing and combing twice a week and clipping every three months.
    A puppy is born black and lightens to blue-gray (ranging from light silver to dark slate). He usually has tan, golden, gray, yellow, or white furnishings. Light coffee brown is also an allowed color but is very rare.
3. His ears hang down. His eyes are dark brown. His nose is black. His teeth meet in a scissors or level bite.
4. Faults: His color should not be more than 20% white.

Health Problems

He is a healthy breed, prone to no real problems.

Cautions when Buying

Don't choose a timid Cesky puppy.

Chinook

Fine for Novice Owners
Good with Children
Large in Size
Thick Medium-Length Coat

	High	Med	Low
EXERCISE REQUIRED		🦴	
TRIMMING/CLIPPING REQUIRED			🦴
AMOUNT OF SHEDDING	🦴		
ACTIVITY INDOORS		🦴	
EASE OF TRAINING		🦴	
SOCIABILITY WITH STRANGERS		🦴	

as they crossed the finish line. The New Hampshire course they had chosen to run on had been named for them long ago—The Chinook Trail—and it seemed appropriate that it wound past Arthur Walden's old farm. This rare breed is a companion and sled dog and a therapy dog for handicapped children.

Temperament

Boisterous as a puppy, he matures in two or three years into a calm, easygoing dog. He is quiet indoors, and although he does better in the suburbs or country where he has space for exercise, he can adapt to the city. Some individuals are mildly friendly with strangers, some are reserved; he is sensibly protective (preferring to guard with his strong body rather than his teeth) and should be accustomed to people at an early age. Most are fine family dogs, but some are rather one-personish. Although independent, he does respond well to firm obedience training. He may refuse commands from family members who have not established leadership over him. Some are a bit nervous. Ongoing exercise and socialization are important with this breed.

History

He is recognized by the Chinook Owners Association. When musher Arthur Walden returned to New Hampshire from the Alaskan gold rush, he brought back Greenland Huskies and crossed them with a Saint Bernard. His most famous lead dog, named "Chinook," which means "warm winds," raced across Canada and accompanied the Byrd Expedition to Antarctica. When Walden retired, the kennel was purchased by an eccentric outdoorsman who refused to sell any breeding stock. The Chinook slowly dwindled toward extinction, but today, stalwart breeders are battling to save this breed with the colorful past and once-bright future. When recently a full team of Chinooks raced in competition for the first time in fifty years, the crowd roared

Physical Features

1. He stands 20–28 inches and weighs 55–70 pounds. Breeders are trying to increase height and weight to more closely match the original powerfully built Chinooks.
2. His coat is coarse with a dense undercoat and should be brushed twice a week regularly; daily when he's shedding. He is shades of tawny, from palomino to reddish gold, often with white or buff trim, and a light or black face mask.
3. His ears may prick up, hang down, or fold back. His eyes may be dark brown or light amber. His nose is black. His teeth meet in a level or scissors bite.
4. Disqualifications: He should not be solid white or particolored.

Health Problems

He is susceptible to hip dysplasia. Buy only from OFA-registered parents.

Cautions when Buying

BE CAREFUL. This is a powerful, independent breed, and if you buy from a poor breeder or raise the dog incorrectly, you could end up with a dominant Chinook. Don't choose the most independent Chinook puppy, and don't choose a timid puppy.

English Shepherd

For Experienced Owners Only
Good with Children
Medium in Size
Feathered Medium-Length Coat

	High	Med	Low
EXERCISE REQUIRED	✦		
TRIMMING/CLIPPING REQUIRED			✦
AMOUNT OF SHEDDING		✦	
ACTIVITY INDOORS		✦	
EASE OF TRAINING	✦	✦	
SOCIABILITY WITH STRANGERS		✦	

Temperament

He is not as high-powered or energetic as an Australian Shepherd or Border Collie, so he often makes a better family pet. He has a calm, steady nature, but he needs too much exercise to live in the city. Although a very willing dog capable of learning almost anything, he does have the independence and good judgment typical of a herding breed. He is reserved and sensibly protective with strangers. Most are fine with other pets if raised with them, but some are aggressive with strange dogs—one of this breed's responsibilities was to chase strays away from his own farm. This is an alert breed who likes to keep his eye on you. He is mature and serious, and he could test owners who do not match his intelligence. Some bark, and some may nip at people's heels as though trying to herd them. Ongoing exercise and activity are important for this breed.

History

He is recognized by the UKC, the International English Shepherd Registry (part of the National Stockdog Registry), and the Animal Research Foundation. Like the Border Collie, he is an old British farm dog brought to American farms by early English settlers. This rare breed is a herding dog and companion.

Physical Features

1. He stands about 20 inches and weighs about 50 pounds.
2. His coat is medium in length with longer hair on his chest, neck, stomach, and legs. He should be brushed and combed twice a week. He is black-and-white, black-and-tan, or sable-and-white.
3. His ears prick up, with the top three-quarters tipping forward. His eyes are dark. His nose is black. His teeth meet in a scissors bite.

Health Problems

He is susceptible to hip dysplasia. Buy only from OFA-registered parents.

Cautions when Buying

Don't choose the most independent English Shepherd puppy.

Glen of Imaal Terrier (Eee-MAHL)

Fine for Novice Owners
Good with Children
Small in Size
Unusual Coat (See Physical Features 2. below)

	High	Med	Low
EXERCISE REQUIRED		●	
TRIMMING/CLIPPING REQUIRED		●	
AMOUNT OF SHEDDING			●
ACTIVITY INDOORS		●	
EASE OF TRAINING		●	
SOCIABILITY WITH STRANGERS			●

Temperament

Although a sturdy, stoic dog with strong working instincts, his rough-and-ready appearance belies his gentle, peaceful, willing disposition. He's active and spirited outdoors but can do well in the city if given enough daily exercise and occasionally taken to a safe, enclosed area to romp. Unlike many terriers, he is a self-reliant dog who does not demand constant attention from his owner. He is cautious and discerning with strangers. Some are scrappy with strange dogs and some will chase pet rabbits and rodents. More willing than many terriers, he responds well to obedience training. He tends to bark a bit in a surprisingly deep, guttural voice that sounds like it's coming from a big dog. He likes to dig.

History

He is recognized by the European FCI, the States Kennel Club, and the Glen of Imaal Terrier Club of North America. From the rugged Glen of Imaal in the Wicklow Mountains of Ireland comes this heavy-boned, crooked-legged hunter of fox, badger, and vermin. Silently drawing his quarry from the hole instead of barking and worrying it to dislodge it, he is quite a game little dog. Fittingly, he was first exhibited at a dog show on St. Patrick's Day in 1933 and has since been shown consistently throughout Great Britain and Western Europe. Enthusiasts feel strongly that he should continue to be hunted in trials so that his working abilities do not become lost in favor of appearance. This rare breed is a companion and vermin hunter.

Physical Features

1. He stands up to 14 inches and weighs 30–35 pounds.
2. His coat is medium in length, harsh, and a bit shaggy, with a dense undercoat, and should be brushed and combed twice a week. Straggly hairs need to be scissored every three months. He is shades of blue brindle or wheaten. Puppies are often born with dark markings that disappear with maturity.
3. His ears are half-pricked or rose ears (the bottom part of the ear is up, but the top part flops over sideways). His eyes are brown. His nose is black. His teeth meet in a scissors or level bite. His tail is docked.
4. Faults: He should not stand over the height specification. His eyes should not be light. His nose should not be pink. He should not have any white in his coat. Adults should not have any black in their coats.

Health Problems

He is a healthy breed, prone to no real problems.

Cautions when Buying

Don't choose a nippy Glen of Imaal puppy.

Havanese

Fine for Novice Owners ✓
Good with Children ✓
Little in Size ✓
Long Coat

	High	Med	Low
EXERCISE REQUIRED ✓			▰
TRIMMING/CLIPPING REQUIRED ✓		▰	
AMOUNT OF SHEDDING ✓			▰
ACTIVITY INDOORS	▰		
EASE OF TRAINING ✓		▰	
SOCIABILITY WITH STRANGERS			▰

Temperament

He is a lively, merry, happy-go-lucky dog with a natural penchant for doing tricks to get attention. Standoffish and sometimes timid with strangers, he should be accustomed to people and noises at an early age. He gets along well with other animals. This gentle little dog is quite responsive to obedience training. Some bark a bit.

History

He is recognized by the Havanese Club of America and the European FCI. He is closely related to the Bichon Frise and the Bolognese. He was the pampered pet of the wealthy ruling class in Havana, Cuba. During the Cuban Revolution, many families fled their homeland and only a few managed to bring their dogs with them, so this breed is being rebuilt from the few dogs that could be found, and he is gradually becoming more popular. He is a companion and also makes a superb circus and trick dog.

Physical Features

1. He stands 8–10 inches and weighs 7–13 pounds.
2. His coat may be wavy or curly and should be brushed and combed every other day or he will be a matted mess. Straggly hairs should be scissored every three months. The most common colors are shades of champagne, white, and golden, but parti-color is also allowed.
3. His ears hang down and are heavily feathered. His eyes are dark. His nose is black. His teeth meet in a scissors or level bite. His tail is plumed over his back.

Health Problems

He is a healthy breed, prone to no real problems.

Cautions when Buying

Don't choose a timid Havanese puppy.

Jack Russell Terrier

Fine for Novice Owners
Good with Older, Considerate Children
Little to Small in Size
Short Coat

	High	Med	Low
EXERCISE REQUIRED		▰	
TRIMMING/CLIPPING REQUIRED			▰
AMOUNT OF SHEDDING		▰	
ACTIVITY INDOORS	▰		
EASE OF TRAINING			▰
SOCIABILITY WITH STRANGERS		▰	

Temperament

He is hardy, bold, and high-spirited, always on top of everything that is going on. He belongs in the suburbs or country where he can get exercise because he can be restless and destructive if confined too much. But don't let him off the leash except in a safe, enclosed area—he is an independent explorer and he will take off in pursuit of anything that moves. He loves to chase balls and play games. He gets along well with strangers, dogs, cats, horses, and livestock, but he will chase rabbits and rodents—and he will catch them. This stubborn little dog needs firm obedience training to keep his energy under control. He is an enthusiastic barker and digger.

History

He is recognized by the JRT Club of America and the JRT Breeders Association of America. He was developed by Parson John Russell of England to run with foxhounds for twenty miles or more. Some British terrier clubs reduced the Reverend's dogs from 12 inches in height all the way down to 9 inches, but this was a corruption that other clubs would not tolerate, so today there are several JRT registries, each with slightly different standards. This is a fairly popular breed, but most breeders do not want AKC recognition for fear working instincts will be neglected in favor of appearance. Enthusiasts feel strongly that he should continue to be hunted in trials so that his working abilities do not become lost as have those of most other terriers. He is a companion and vermin hunter.

Physical Features

1. He stands 12–14 inches and weighs 10–15 pounds.
2. He has a hard coat, which needs only a quick brushing once a week. There is also a less-common rough-coated or broken-coated Jack Russell Terrier, with a wiry coat. He is mostly white with a few black and/or tan markings.
3. His ears fold forward. His eyes are dark. His nose is black. His teeth meet in a scissors bite. His tail is docked.
4. Faults: His eyes should not be yellow. His body should not be too heavily marked with color patches.
5. Disqualifications: His markings should not be brindle. His ears should not prick up. His nose should not be brown or white. His teeth should not meet in a badly overshot or undershot bite.

Health Problems

He is susceptible to a congenital dwarfism where a puppy is born with a normal head and body but with short, warped legs.

Cautions when Buying

Be sure you are buying a registered JRT, because some dishonest or ignorant breeders are selling white-and-patched mixed breeds as JRTs. Don't choose the boldest, most independent, or most energetic JRT puppy.

Kyi-Leo (KYE-LEO)

Fine for Novice Owners
Good with Older, Considerate Children
Little in Size
Long Coat

	High	Med	Low
EXERCISE REQUIRED			✦
TRIMMING/CLIPPING REQUIRED			✦
AMOUNT OF SHEDDING		✦	
ACTIVITY INDOORS	✦	✦	
EASE OF TRAINING		✦	
SOCIABILITY WITH STRANGERS			✦

Temperament

He is bright, lively, agile, and able to grasp objects with his front paws. Extremely people-oriented and playful, he likes personal attention and companionship. He is reserved and sometimes timid with strangers and should be accustomed to people and noises at an early age. He gets along well with other animals. He is sensitive but also stubborn, so he needs gentle but firm obedience training. He barks a bit.

History

He is recognized by the Kyi-Leo Club of America. He was developed by the Linn family in the San Francisco Bay area from Lhasa Apso and Maltese crosses. *Kyi* is Tibetan for dog (crediting the Lhasa) and *Leo* is Latin for lion (crediting the Maltese). This rare breed is a companion.

Physical Features

1. He stands 9–11 inches and weighs 8–12 pounds.
2. His coat is silky and parted down the center of his back, and should be brushed and combed every other day or he will be a matted mess. Puppies' coats take three to four years to reach full length. He is usually black-and-white with black ears and black markings around his eyes and sometimes brownish shadings on his ears and above his eyes. Sometimes the black lightens to silvery gray. Other colors are acceptable.
3. His ears hang down and are heavily feathered. His eyes are dark. His nose is black. His teeth meet in a scissors, level, or slightly undershot bite. His tail plumes over his back when he is excited.

Health Problems

He is susceptible to slipped stifle (a joint disorder that may require surgery) and flea allergies.

Cautions when Buying

Don't choose a timid Kyi-Leo puppy.

Leonberger

For Experienced Owners Only
Good with Children
Giant in Size
Thick Medium-Length Coat

	High	Med	Low
EXERCISE REQUIRED	▰		
TRIMMING/CLIPPING REQUIRED			▰
AMOUNT OF SHEDDING		▰	
ACTIVITY INDOORS			▰
EASE OF TRAINING		▰	
SOCIABILITY WITH STRANGERS		▰	

Temperament

He is a noble, gentle giant—easygoing, patient, reliable, somewhat serious with a sociable, generous nature. Although quiet and calm indoors, he is enthusiastic and energetic outdoors, so he belongs in the suburbs or country with someone who will take him out regularly. Unlike some giant breeds, he is not lethargic or cumbersome but very agile. He's mildly friendly with strangers, and although protective instincts develop at maturity (three years) and he becomes more discriminating, he is never aggressive. He's fine with other animals if raised with them, but some males can be aggressive with strange male dogs. A willing dog, he responds well to firm obedience training. He likes to swim, track, and pull carts and sleds. He can refuse commands from family members who have not established leadership over him. He does *not* drool.

History

He is recognized by the Leonberger Club of America and the European FCI. The official crest of the city of Leonberg, Germany included a lion, so the patriotic mayor crossed Saint Bernards, Great Pyrenees, and Newfoundlands to develop a dog who resembled a lion. The breed became popular with the rich and famous: the Empress of Austria, Napoleon III, the German composer Richard Wagner. Few Leonbergers survived the famine and bombings of World War II, and the breed is still rare in the United States. He is a companion and has been used for backpacking, pulling carts and sleds, earthquake and water rescue, livestock protection, and Schutzhund competition.

Physical Features

1. Males stand 29–32 inches and weigh 110–150 pounds. Females stand 26–30 inches and weigh 90–120 pounds.
2. His coat is slightly wavy with a dense undercoat and should be brushed twice a week. A neck mane or ruff develops between two and three years of age. He is lion-colored (golden yellow to reddish brown), although some hairs may be dark-tipped. Preferably, he has a dark mask on his face. A little white on his chest and toe tips is allowed.
3. His ears hang down. His eyes are brown with a friendly expression. His nose is black. His teeth meet in a scissors bite. His feet are usually webbed to aid in swimming.
4. Faults: His eyes should not be yellow.
5. Disqualifications: His nose or lips should not be brown.

Health Problems

He is a short-lived breed (ten years) and susceptible to hip dysplasia, osteochondritis (bone disease), and eyelid abnormalities. Buy only from OFA-registered parents.

Cautions when Buying

BE CAREFUL. This is a powerful breed, and if you buy from a poor breeder or raise the dog incorrectly you could end up with an aggressive boss-dog Leonberger. Don't choose the boldest, most energetic, or most independent Leonberger puppy, and don't choose a timid puppy.

Louisiana Catahoula Leopard Dog

For Experienced Owners Only
Good with Children if Raised with Children
Large in Size
Smooth Coat

	High	Med	Low
EXERCISE REQUIRED	▪		
TRIMMING/CLIPPING REQUIRED			▪
AMOUNT OF SHEDDING		▪	
ACTIVITY INDOORS		▪	
EASE OF TRAINING		▪	
SOCIABILITY WITH STRANGERS			▪

Temperament

Although usually calm indoors, he's so energetic outdoors and in need of so much hard exercise that he is completely unsuited to the city. Confinement makes him noisy, hyperactive, and prone to behavior problems. This reserved, aggressive home guardian must be accustomed to people at an early age. He's okay with other pets if raised with them, but many are aggressive with strange dogs. This is a strong-willed, independent breed, but if his trainer is experienced and works him in obedience early, he is capable of learning almost anything. He loves to swim. He can guard objects and refuse commands from family members who have not established leadership over him. Ongoing exercise, socialization, and activity are essential with a Catahoula.

History

He is recognized by the National Association of Louisiana Catahoulas. He is the state dog of Louisiana. He was developed in the Catahoula Lakes region of the Deep South, a blend of red wolf and fierce dogs used by the explorer DeSoto to conquer the Indians of that region. The strikingly colored progeny were adopted by white settlers and eventually went to Texas, where the cowboys called them Leopard Dogs. This versatile working dog has an enviable reputation with wild cattle and Brahma bulls. He is a silent air-scenting trailer of deer, and a superior coon and wild-hog hunter. He is also a guardian and Schutzhund dog. *Catahoula* means "beautiful clear water," which may also be a tribute to the breed's eerie translucent blue eyes.

The "Cat" is rare outside the bayous and swamps of the South.

Physical Features

1. Males stand 22–26 inches; females stand 20–24 inches. He weighs 50–95 pounds.
2. His hard coat needs only a quick brushing once a week. The most treasured colors are the Blue Leopard or Red Leopard: a spotted, patched, mottled mix of blue/gray/black/liver/red/white. Solid black, red, chocolate, or yellow are also allowed, as is brindle.
3. His ears may hang down or fold backward. Blue-white ("glass") eyes are preferred, but brown or green is allowed. His teeth meet in a scissors or level bite. His feet are webbed for swimming and for traction in swamp and mud.

Health Problems

Catahoulas who are mostly white are susceptible to inherited deafness. Test hearing by whistling or clapping.

Cautions when Buying

BE CAREFUL. This is a powerful, independent breed, and if you buy from a poor breeder or raise the dog incorrectly, you could end up with a very aggressive Catahoula. Don't choose the boldest, most independent, or most energetic Cat puppy, and don't choose a timid puppy.

Löwchen (Little Lion Dog)

Fine for Novice Owners
Good with Older, Considerate Children
Little to Small in Size
Long Coat

	High	Med	Low
EXERCISE REQUIRED			
TRIMMING/CLIPPING REQUIRED	✶		
AMOUNT OF SHEDDING		✶	
ACTIVITY INDOORS	✶		
EASE OF TRAINING		✶	
SOCIABILITY WITH STRANGERS		✶	

Temperament

Some are high-spirited and outgoing, some are calmer, some are a bit timid or nervous. Most are lively but also sensible and low-key. He is much sturdier than he looks, needs some exercise, and loves to run and play. Some are friendly with strangers, some are reserved, some are timid, so he should be accustomed to people and noises at an early age. He gets along well with other animals. He's willing to please and responds quite well to obedience training. Some are barkers or diggers.

History

He is recognized by the Little Lion Dog Club of America and the European FCI. In medieval times, a knight killed in battle would have a lion emblazoned on his tomb, but if killed in a less-valiant endeavor, he would have a Little Lion Dog, so named only for his clipped appearance, not for his courage. He was the companion of Florentine nobility around the Mediterranean and is probably related to the Bichon Frise, Bolognese, and Havanese. He has been portrayed in tapestries, the woodcuts of Albrecht Dürer, and the oil paintings of Goya. Nearly extinct after WWII, the breed was kept alive by an eccentric elderly woman who scoured the war-torn streets for Little Lion Dog representatives. When the first Löwchen touched down on American soil in 1971, the press was waiting on the airfield tarmac. This rare breed is a companion.

Physical Features

1. He stands 9–15 inches and weighs 10–18 pounds.
2. His harsh wavy coat grows naturally in a rough lion pattern with a neck ruff and is clipped to emphasize these lines. He should be brushed and combed every other day or he will be a matted mess. He may be any color or color combination, including golden, brown, apricot, parti-color, and others.
3. His ears hang down. His eyes are dark but may be lighter in brown and apricot dogs. His nose is usually black but is brown on brown dogs. His teeth meet in a scissors or level bite. His tail is plumed over his back.

Health Problems

He is a healthy breed, prone to no real problems.

Cautions when Buying

Don't choose a timid Löwchen puppy.

Norwegian Buhund (BOO-hund)

Fine for Novice Owners
Good with Children
Medium in Size
Thick Medium-Length Coat

	High	Med	Low
EXERCISE REQUIRED		⟫	
TRIMMING/CLIPPING REQUIRED			⟫
AMOUNT OF SHEDDING		⟫	
ACTIVITY INDOORS		⟫	
EASE OF TRAINING		⟫	
SOCIABILITY WITH STRANGERS		⟫	

Temperament

He is extremely alert, happy, and outgoing, with keen senses and very mobile ears. Although high-spirited, energetic, and needing plenty of exercise, he is less boisterous and more family-oriented than are many other spitz breeds. Some are friendly with strangers, some are watchful. He gets along well with other pets if raised with them. Mildly independent but usually willing to please, he responds well to obedience training. He's light on his feet and excels in agility competition (an obstacle course for dogs). This adaptable dog makes a pleasant traveling companion. He is a barker with a rapid, high-pitched bark. He may nip at people's heels as though trying to herd them.

History

He is recognized by the Norwegian Buhund Club of America. Very popular on Scandinavian farms, he is an adept herder and watchdog and occasionally hunts small game. *Bu-hund* means homestead dog. This very old breed served as the companion of mountain shepherds who lived in crude huts while their stock grazed. Very rare in the United States, he is a companion and herding dog.

Physical Features

1. He stands 15–17 inches and weighs 25–40 pounds.
2. His coat is harsh with a dense undercoat and needs brushing twice a week. He is usually wheaten (cream, apricot, or light brown), often with black-tipped hairs and a black mask. He may also be black, with a white blaze, chest, paws, and neck ring allowed.
3. His ears prick up. His eyes are dark brown. His nose is black. His teeth meet in a scissors bite. His tail curls along the middle of his back.
4. Faults: His nose should not be brown or pink. His eyes should not be light colored. His ears should not hang down. His tail should not hang down or be loosely curled when he is in motion. His coat should not be long or wavy. He should not be any color other than wheaten or black.

Health Problems

He is susceptible to hip dysplasia and PRA, and good breeders are routinely X-raying hips and checking eyes.

Cautions when Buying

Don't choose the most energetic or independent Buhund puppy or a timid puppy.

Norwegian Lundehund (Puffin Dog)

Fine for Novice Owners
Good with Older, Considerate Children
Small in Size
Short Coat

	High	Med	Low
EXERCISE REQUIRED		▰	
TRIMMING/CLIPPING REQUIRED			▰
AMOUNT OF SHEDDING		▰	
ACTIVITY INDOORS		▰	
EASE OF TRAINING			▰
SOCIABILITY WITH STRANGERS		▰	

Temperament

Indoors he is calm and easygoing; he will come over for attention, then be content to lie down. Outdoors he is lively and playful and likes to carry things around in his mouth. Although quick to alert and bark, he is a gentle, totally nonaggressive dog. He likes people and other animals, but adolescents and females can be timid, so he should be accustomed to people, noises, and new situations at an early age. Obedience training must be gentle: He is obstinate and independent but also highly sensitive. He is exceptionally agile, with a springy gait. This breed retains many primitive instincts, such as digging holes to hide food.

History

He is recognized by the Norwegian Puffin Dog Club of America. It has been suggested that he is the only dog to have survived from before the last ice age. Isolated for centuries on rugged, rocky Norwegian islands, he hunted the puffin (*lunde*) bird for its meat and downy feathers. The only canine with six toes and the ability to grasp with them, he is lightning swift and surefooted as he scales impossibly steep, slippery rock walls to reach roosting dens. Squeezed into tight crevasses, he can bend his neck backward until his nose touches his spine—the reindeer is the only other mammal with this neck structure—and he can flatten his flexible shoulders and front legs into a bearskin-rug position. He can also close his prick ears to protect them from the moisture and dirt of the dank caves. However, a ban on puffin hunting and an outbreak of distemper nearly wiped out the breed, and although Norwegians are slowly rebuild-

ing it from six remaining dogs, it is still rare and unique. He is a companion.

Physical Features

1. He stands 13–15 inches and weighs 12–15 pounds.
2. His coat is rough and should be brushed once a week. He's usually reddish brown or yellowish brown, with black-tipped hairs in adults. He may also be black or gray. He always has white markings.
3. His ears prick up. His eyes are light brown. His teeth meet in a scissors, level, or slightly undershot bite. His tail usually rings over his back when he is alert. He does not have any premolar teeth.

Health Problems

He is extremely susceptible to severe intestinal problems, so he must be fed a high-protein, low-fat dog food. Never feed him greasy table scraps.

Cautions when Buying

Don't choose a timid Lundehund puppy.

Nova Scotia Duck Tolling Retriever

Fine for Novice Owners
Good with Children
Medium in Size
Feathered Medium-Length Coat

	High	Med	Low
EXERCISE REQUIRED	●		
TRIMMING/CLIPPING REQUIRED		●	
AMOUNT OF SHEDDING		●	
ACTIVITY INDOORS	●		
EASE OF TRAINING		●	
SOCIABILITY WITH STRANGERS		●	

Temperament

In some ways—his good nature and wagging tail—he's similar to the Golden Retriever he also resembles physically, but in other ways he's different. Easygoing but also high-spirited and playful, he does best in the suburbs or country where he can get exercise. Some are friendly with strangers, some are reserved, some are timid, so he should be accustomed to people and noises at an early age. He's usually fine with other animals—he may chase the family cat for fun but seldom means him any harm. Although he generally wants to please, he can be stubborn and clever, and sometimes shows his sense of humor when disobeying. He needs firm (but not harsh) obedience training. He whines and makes a whistling sound when excited. Some are diggers and some youngsters chew destructively.

History

He is recognized by the Canadian Kennel Club. He was probably developed from retrievers, setters, spaniels, and possibly farm Collies. He was once called the Little River Duck Dog and the Red Decoy Dog. Unlike most retrievers, he does not simply sit beside the gunner and wait for the birds to fly into range and be shot down. His job is to run up and down the shoreline chasing sticks and furiously wagging his tail to lure curious ducks into gun range. This is called "tolling," and is the same clever method used by the fox to lure prey within his reach. This breed is rare in the United States. He is a companion.

Physical Features

1. He stands 18–20 inches and weighs 35–50 pounds.
2. His coat is soft and wavy, with longer hair on his chest, stomach, legs, and tail, and should be brushed and combed twice a week; straggly hairs need to be scissored every three months. He is shades of red or orange, usually with a white face blaze, white chest, white feet, and/or white tail tip.
3. His ears hang down. His eyes are brown or amber (brownish yellow) with a sad expression. His nose is black, brown, or flesh colored. His teeth meet in a scissors bite. His feet are webbed to aid in swimming.
4. Faults: His nose should not be bright pink.
5. Disqualifications: He should not have any white on his shoulders, around his ears, on the back of his neck, or across his back or hips. He should not be any color other than red or orange. He should not have any black or gray hairs. His feet should not be nonwebbed. His bite should not be overshot more than ⅛ inch or undershot. His nose should not be spotted or patched.

Health Problems

He is susceptible to hip dysplasia and PRA. Buy only from OFA- and CERF-registered parents.

Cautions when Buying

Don't choose the most independent Toller puppy, and don't choose a timid puppy.

Polish Owczarek Nizinny Sheepdog (Ov-CHA-rek Nee-ZHEE-nee)

For Experienced Owners Only
Good with Children
Medium in Size
Long Coat

	High	Med	Low
EXERCISE REQUIRED		●	
TRIMMING/CLIPPING REQUIRED			●
AMOUNT OF SHEDDING		●	
ACTIVITY INDOORS		●	
EASE OF TRAINING		●	
SOCIABILITY WITH STRANGERS			●

Temperament

He is calm, confident, sensible—an easy keeper. He does well in the city, but he's an active, robust dog who does need daily exercise. Reserved and watchful with strangers, he should be accustomed to people at an early age. He can be scrappy with strange dogs. He's adaptable and makes an excellent traveling companion. He is responsive to firm, fair obedience training, but he is also clever, strong-willed, and likes to think for himself and make his own decisions. This is a serious, responsible dog who could bite if provoked. He may nip at people's heels as though trying to herd them.

History

He is recognized by the States Kennel Club, the European FCI, and the American Polish Owczarek Nizinny Club. The first native Polish breed to be introduced to the United States, this very old livestock herder is descended from the Hungarian Puli and sixteenth-century Hun herding breeds. His name is sometimes affectionately shortened to PONS, and he is also sometimes called the Polish Lowland Sheepdog because he worked on the lowland plains. This rare breed is a companion and herding dog, and has proven to be a good therapy dog for children, seniors, and the handicapped.

Physical Features

1. He stands 16–20 inches and weighs 30–50 pounds.
2. His coat is shaggy, sometimes slightly wavy, covers his eyes, and should be brushed and combed every other day or he will be a matted mess. He may be any color or pattern of colors.
3. His ears hang down. His eyes are hazel or brown. His nose is dark. His teeth meet in a level bite. His tail is a natural bobtail or docked short.
4. Faults: His nose should not be pink. His eyes should not be light yellow.

Health Problems

He is a healthy breed, prone to no real problems.

Cautions when Buying

Don't choose the most independent PONS puppy.

Shiba (SHEE-ba)

Fine for Novice Owners
Good with Children
Small in Size
Thick Medium-Length Coat

	High	Med	Low
EXERCISE REQUIRED		🦴	
TRIMMING/CLIPPING REQUIRED			🦴
AMOUNT OF SHEDDING		🦴	
ACTIVITY INDOORS		🦴	
EASE OF TRAINING		🦴	
SOCIABILITY WITH STRANGERS		🦴	

Temperament

He is spirited, bold, and alert, but indoors, while lively, he's quiet and clean and not a nuisance underfoot. He is a big dog in a small package—hardy and capable, he's not a lapdog. Most are cautious with strangers, but some are friendly. He is fine with other dogs and cats if raised with them, but this agile, quick breed should be watched around very small pets like birds, rabbits, and rodents. He's independent but responds well to firm, fair obedience training. He's adaptable and makes an excellent traveling companion. Because of his foxy appearance, he should be carefully supervised during hunting seasons in your area.

History

He is recognized by the Japan Kennel Club and three Shiba registries in the United States. This very old Japanese breed is extremely popular in Japan. His original name, *Shiba Inu*, means "little brushwood dog," in reference to the dense low bushes of Japan through which he hunted birds and small game. Today "Inu" has been largely dropped. This versatile breed is becoming more popular in the United States. He is a companion.

Physical Features

1. Males stand 15–16 inches, females stand 13¾–15 inches. He weighs about 15–28 pounds.
2. His coat is velvety and should be brushed twice a week. He is red, red sesame (red with a sparse black overlay), or black with tan markings above his eyes, on his lower jaw, cheeks, chest, legs, and tail. White markings are required on the sides of his muzzle (not across the top) and on his cheeks, throat, belly, and chest (as far back as his shoulder). White markings are allowed on his legs, the tip of his tail, and above his eyes.
3. His ears prick up and incline forward. His eyes are dark brown. His nose is black. His teeth meet in a scissors bite. His tail curls (sometimes double) over his back.
4. Faults: His eyes should not be yellow or blue-gray. His nose should not be any color other than solid black. His tongue should not be black-spotted. His teeth should not meet in a level bite. He should not have tan or white markings extending all over his head, shoulders, or front legs. A red sesame should not have a black saddle.
5. Disqualifications: He should not stand more than ½ inch over the height specification. His ears should not be hanging or tipped. He should not have more than four missing teeth. His tail should not be incapable of curling over his back. He should not be missing his white markings. His teeth should not meet in an overshot or undershot bite. He should not be any color than those allowed, such as cream, white, brindle, or pinto.

Health Problems

He is susceptible to hip dysplasia and PRA. Buy only from OFA- and CERF-registered parents.

Cautions when Buying

Don't choose the boldest or most independent Shiba puppy, and don't choose a timid puppy.

Swedish Vallhund

Fine for Novice Owners
Good with Children
Small in Size
Short Coat

	High	Med	Low
EXERCISE REQUIRED		●	
TRIMMING/CLIPPING REQUIRED			●
AMOUNT OF SHEDDING		●	
ACTIVITY INDOORS	●		
EASE OF TRAINING		●	
SOCIABILITY WITH STRANGERS	●		

Temperament

He is a bright, busy, steady-tempered little dog who does fine in the city if exercised enough. He is friendly with strangers and fine with other animals, especially enjoying other Vallhunds. Although slightly independent and used to thinking for himself, he responds quite well to obedience training. This adaptable and hardy little dog makes an excellent traveling companion. He is a barker, and he may nip at people's heels as though trying to herd them.

History

He is recognized by the United Kennel Club and the Swedish Vallhund Enthusiasts of America. This rugged little spitz is a cattle herder and farm dog brought to Sweden by the early Vikings. Whether he has crossed paths with the similar-looking Pembroke Welsh Corgi is not known; the Vallhund is leggier and not as long-bodied as the Corgi, but both breeds have the same working style: They dart in and nip the cattle on their hocks, then drop quickly to the ground to avoid the ensuing kick. *Vallhund* means "forest dog." The breed's Swedish name is *Västgötaspets*, meaning "a spitz breed from the Swedish province of Västergötland." He is extremely rare in the United States. He is a companion, a herding dog, a vermin hunter, and is also proving successful in therapy programs for children, seniors, and the handicapped.

Physical Features

1. He stands 12–13 inches and weighs 20–35 pounds.
2. His coat is coarse with a dense undercoat and should be brushed once a week. He is blended shades of gray, brown, yellow, and/or reddish, with darker hairs on his back, neck, and sides, and lighter hairs or white hairs on his muzzle, throat, chest, belly, hindquarters, and lower legs.
3. His ears prick up. His eyes are dark brown. His nose is black. His teeth meet in a scissors or level bite. His tail is a natural bobtail or docked under 4 inches.
4. Faults: White markings should not cover more than 20% of his body.

Health Problems

He is a long-lived (fifteen years) healthy breed, prone to no real problems.

Cautions when Buying

Don't choose the most independent or most energetic Vallhund puppy.

Toy Fox Terrier

Fine for Novice Owners
Good with Older, Considerate Children
Little in Size
Smooth Coat

	High	Med	Low
EXERCISE REQUIRED			🦴
TRIMMING/CLIPPING REQUIRED			🦴
AMOUNT OF SHEDDING		🦴	
ACTIVITY INDOORS	🦴		
EASE OF TRAINING		🦴	
SOCIABILITY WITH STRANGERS		🦴	

Temperament

He is spirited, playful, and bold for his size, although some are a bit high-strung. He is an ideal apartment dog but does like his walks. He likes to participate in games and activities, needs attention, and has a special affinity for the elderly. Some are friendly with strangers, some are reserved, some are timid, so he should be accustomed to people at an early age. He can be scrappy with other animals. He generally wants to please but can be stubborn and needs a firm hand. He doesn't enjoy learning obedience exercises as much as he enjoys learning tricks. Some bark, some dig, and some chew destructively if left alone too much.

History

He is not recognized by the AKC, but by the United Kennel Club. He was bred down from the Smooth Fox Terrier and is sometimes called an Amertoy (American Toy Fox Terrier). Many older folks remember having these dogs around the farm when they were growing up. He is a companion, a great mouser, and so alert that he is sometimes used in the Hearing Ear (Dogs for the Deaf) program.

Physical Features

1. He stands up to 10 inches and weighs 4–7 pounds.
2. He has a satiny coat, which needs only a quick brushing once a week. Preferably, he is white with black spots and tan trim. Sometimes he is white-and-tan or white-and-black. His head is black with tan dots over his eyes and on the sides of his muzzle, and he may have a white blaze.
3. His ears prick up. His eyes are dark. His nose is black. His teeth meet in a scissors bite. His tail is docked.
4. Faults: His nose should not be brown or pink-spotted. He should not be a solid color. His markings should not be chocolate.

Health Problems

He is susceptible to some eye problems and skin conditions. He is sensitive to cold and should wear a sweater outdoors in very cold weather.

Cautions when Buying

Be sure you are buying a UKC-registered Toy Fox Terrier, because some dishonest or ignorant breeders are selling patched mixed breeds as Toy Fox Terriers. Don't choose the most independent, scrappy, or excitable Toy Fox, and don't choose a timid puppy.

STEP THREE

Choosing the Right
Breeder

6

Why Should You Buy from a Breeder?

Sources for purebred puppies include good breeders, poor breeders, pet shops, individuals (friends, neighbors, coworkers), animal shelters, and breed rescue leagues.

Do you wonder why I've included the ominous-sounding "poor breeders" as a source for a purebred puppy? It's because there are so many of them and you must learn how to recognize them so you can *avoid* them.

First, let's talk about good breeders. A good breeder owns a *purebred* and *registered* female. Wait—aren't purebred and registered the same thing? Not at all. A dog can be purebred without having registration papers. You've already learned that being purebred means that a dog and all his relatives have the same fixed set of genes developed over a long period of time. "Papers" or "no papers" has no effect on those fixed genes.

"Papers" simply means that a registry organization, such as the AKC, has followed a chain of paperwork sent in by the owners of your dog's parents, grandparents, and other progenitors, and has formally concluded that since all your dog's relatives were purebred, your dog must be, too. For a fee, they'll add your dog's name to the chain and send you a registration certificate (his papers).

Thus, registration is a purely mechanical process that operates on the honor system: the AKC takes everyone's word for it that Barney and Jessie were really the parents of Solomon. To the AKC, your dog is one number in a long line of numbers; they've never seen him, his parents, or his grandparents, thus they cannot and do not judge any dog's quality. A registration certificate (or the lack of one) has nothing at all to do with a dog's quality.

What about a *pedigree*? Isn't that something special? Again, not at all. A pedigree is a family tree, a list of ancestors—nothing more and nothing less. Theoretically, every dog—purebred and mixed—has a pedigree, because every dog, like every person, has a family tree. If you tracked down the names of your mixed breed's parents, grandparents, and so on, your mixed breed would have a pedigree, too.

Realistically, though, we do usually speak of a pedigree as the family tree of a registered purebred where all the names belong to other registered purebreds. But just because the dogs are registered purebreds does not preclude them from being Black Barts. It's the actual *dogs* behind the names that are important, not the list of names. So the mere existence of a pedigree, like the mere existence of registration papers, says nothing at all about a dog's quality.

Quality: We keep saying that. What's quality? A purebred dog may be of good quality, poor quality, or something in between. A purebred of good quality has a sound temperament: He doesn't bite indiscriminately, he doesn't flee indiscriminately, and he gets along reasonably well with the rest of the world. A purebred of good quality doesn't have (or carry) any congenital medical disorders. A purebred of good quality resembles his breed and acts like his breed—he fits the *standard* for his breed.

What's a standard? Each breed has a written standard of perfection that describes a perfect dog of that breed—exactly what physical and temperamental characteristics he should have that would best equip him to do the work for which he was developed.

Breeds look and act as they do for good reason. Those from desert climates wear short coats; those from cold climates wear thick coats. Those that chased prey across open terrain are swift and long-legged; those that chased vermin into underground dens have short legs and a tenacious personality. Because each breed was developed for a purpose, certain characteristics were found to be best for accomplishing that purpose. Those characteristics are listed in each standard.

So a dog who fits the standard for his breed is said to be a good representative of his breed. He has most of the desirable characteristics, he does not have too many *faults* (undesirable characteristics), and he has no *disqualifications* (extremely undesirable characteristics). A dog is compared to the standard for his breed to see how close he comes to the good traits, and how few of the faults and disqualifications he has.

Why are the standards important? The standards were written so breeders would have something concrete to aim at. With no organized guidelines, dogs would be bred willy-nilly, with countless personal prejudices determining how a breed looks and acts. It would be too easy to change a breed into something unrecognizable. Those who love purebred dogs try to preserve them as they originally were intended to be; thus, the standards exist as quality control.

A good breeder abides by this quality control. His dogs are not only purebred and registered, but also good tempered, healthy, X-rayed to be sure they're clear of congenital problems, and fine representatives of their breed. A good breeder breeds a quality female to a quality male and takes extra-special care of the pregnant female and the newborn puppies. Finally, a good breeder searches for the best home for each puppy, asking probing questions of prospective buyers about their ability to properly care for the puppy. You'll soon be a buyer yourself, so don't be put off by these questions. You must realize that you're not buying a piece of merchandise but a little living creature who has been carefully bred and lovingly raised by a concerned, dedicated breeder. This is exactly the type of good breeder from whom you want to buy your puppy.

Now for the flip side: What is a *poor breeder?* Unfortunately, poor breeders don't wear signs around their necks identifying them as such, though there are plenty of clues that should enable you to spot them from miles away.

Registry organizations say that "a breeder is the person who owns the mother of a litter." That means that if Mrs. Dipp's fat female Labrador, who runs loose, mates with a male Labrador who also runs loose, Mrs. Dipp is a breeder, even though she was blissfully unaware that her chubby Lab was even in the family way until the puppies popped out.

Webster's dictionary says that "a breeder is a person who causes animals to reproduce by controlled mating." That's what Biff did when he decided to create great guard dogs by letting his female Rottweiler "fool around" with a male Rottweiler. Two months later, Breeder Biff was delighted to welcome seven cute puppies. Four months after *that*, Breeder Biff was not so delighted when he had to truck seven gawky puppies to the animal shelter because nobody wanted a Rottweiler puppy who growled.

How about the Greedybucks? They breed their Saint Bernards together every six months and hand over a puppy to anyone who pays a hundred dollars. The Greedybucks have already sold one puppy to a man who sponsors dogfights and another to a research lab. Do the Greedybucks sound like people who abide by the Saint Bernard standard? Do they sound like people who have even *heard* of the Saint Bernard standard?

How about the Snowsnob Kennels? They deliberately breed white German Shepherds, even though white is such a serious deviation from the AKC German Shepherd Standard that it's a disqualification. (Remember, though, that the AKC doesn't consider quality when registering dogs; thus, white Shepherds can be registered because they are purebred. But white is such an undesirable color for Shepherds that white dogs cannot compete in the show ring.)

Why not? What's wrong with white Shep-

herds? Rest assured, disqualifications like white color weren't just snatched out of the air. The German Shepherd was developed to help shepherds herd their flocks, and it's difficult for a shepherd to distinguish between a white dog and white sheep. It's equally difficult for the sheep to distinguish between a white dog and their white friends; thus, sheep tend to behave better for darker dogs. Shepherds are also police dogs and military dogs, and since white stands out, it puts the dog's life—and that of his master—in danger, hence guard-dog breeds are usually dark as well.

But suppose your pet Shepherd isn't going to herd sheep or stalk midnight trespassers. What's the harm if he's white? The harm may be in his genes. Since most breeders *are* conscientious about breeding quality Shepherds, there aren't that many whites around. Thus, an all-white breeding program usually includes any and all white Shepherds that the breeder can find, regardless of their overall quality. Any time a breeder focuses on a single factor, he must limit himself to dogs who can deliver that factor, even if those dogs are deficient in the more important factors of temperament, health, and overall structure.

Mrs. Dipp, Breeder Biff, the Greedybucks, and Snowsnob Kennels are all poor breeders. In other words, *poor breeders* are those who, whether ignorantly or deliberately, consistently produce puppies who may indeed be purebred and registered but who may be deficient in the critical areas of temperament, health, and structure.

The conclusion—to buy from a good breeder—seems obvious, but one caveat should be noted: Good breeders occasionally produce poor puppies, too. Since invisible gene combinations can't always be predicted with *complete* certainty, sometimes a breeder can conscientiously do everything right and the imprecise science of genetics simply does something unexpected.

So if you're looking at a "faulty" or "disqualified" puppy produced by a good breeder, consider: If the problem is with his temperament or health, don't buy. But if the problem is with his appearance and if you're not interested in the show ring, you may decide to overlook some cosmetic faults in an otherwise good puppy from a good breeder. Cosmetic faults certainly don't detract from a puppy's ability to be a wonderful companion; they are cause for concern only in that they may be an *indication* that you're dealing with a poor breeder.

When a breeder deliberately produces nonstandard dogs, or couldn't care less that he's producing nonstandard dogs, or doesn't even *know* that he's producing nonstandard dogs, avoid him. A breeder who has slacked off on his choice of parents as far as their appearance goes, may have slacked off on their temperament and health as well, and the sad or tragic results will eventually show up in the puppy you bring home.

Thus, it is not really faults or disqualifications that are a threat to any puppy's potential as a good pet, but the fact that such faults and disqualifications *may be* indicative of ignorance or carelessness that could well have carried over into the more important areas of temperament, health, and structure.

This is precisely why you should avoid *pet shops*. Most pet-shop puppies barely resemble their breeds. Dedicated breeders who have been involved with their beloved breeds for years and years cringe when they see a representative in a pet-shop window.

Where do pet-shop puppies come from? Definitely not from good breeders. Pet-shop puppies come from the poor breeders who couldn't be bothered confining their females, or choosing a male carefully, or raising the puppies properly, or searching for suitable homes.

Pet-shop puppies also come from commercial puppy mills or puppy farms. Based primarily in the Midwest, puppy mills operate this way: Females of all the popular breeds are kept in cramped cages in crowded backyards and bred at every heat period to any male of the same breed—sometimes not even of the same breed.

But then the puppies won't be purebred, you say. You're right. So they can't be registered, you say. You're wrong.

Let's say a puppy-mill operator has a registered Beagle male who dies. The mill operator doesn't wring his hands in despair that he's lost his Beagle stud—he simply uses a Beagle-mix in the Beagle's place, but he continues to send the Beagle's registration number to the AKC whenever he registers a litter. Remember, registration is a mechanical process; the AKC never sees the dogs they register. Registration papers are only as honest as the person reporting the information.

I once had to tell an owner that her "Doberman" puppy, complete with registration papers from the pet shop, was not a purebred Doberman, but a Doberman/Shepherd cross. The pet shop probably hadn't known that the papers were fraudulent, although someone in the pet shop should have noticed *this* puppy. But pet shops who buy from puppy mills know that they're dealing with shady people and that frauds can occur.

What other frauds can mill operators perpetrate? Sometimes a mill operator buys a male who has done well enough in the show ring to gain his championship. This dog, say a Pembroke Welsh Corgi, may have been sold by his breeder because, although his own structure was good enough for a successful show career, tucked in with his good genes were some poor genes that he kept passing on to his offspring. Remember when we discussed the dog whose own ears were properly pricked but whose genes were one-prick-one-hanging, and he kept passing on the hanging-ears gene? You learned that what's on the outside is not always what's on the inside.

So a mill operator buys a great-looking Corgi with not-so-great genes. The title of "Champion" on the pedigree looks wonderful to mill operators, pet-shop proprietors, and the un-suspecting public. The poor genes passed on to the puppies may not be revealed until their ears refuse to prick up, their coats grow long like a Collie's, and they develop hip dysplasia. The local veterinarian hears: "I don't understand it—his father was a *champion!*"

Besides bad parentage, the mill puppy's early care is horrible. Whisked away from his mother and littermates at four to six weeks of age and shipped to the pet shop, he is deprived of many early lessons that can only be learned from other dogs. Puppies taken away this early are often unable to get along with other dogs—and sometimes with people, as well. Then the local obedience instructor hears: "I don't under-stand it—he was the cutest one in the pet-shop window!"

By now I hope it's clear to you that when you buy one of those cute puppies in the pet-shop window, you often buy more than the puppy. You buy a multitude of budding physical and behavioral problems created by lack of early care, the trauma of shipping, and poor genes passed on by poor parents whom you never get a chance to see and evaluate. Obedience instruc-tors and veterinarians across the country will readily attest to the disproportionate number of problems that develop in pet-shop puppies as they grow.

In addition, many pet-shop clerks don't find out what their customers are really looking for in a dog. If they do ask, they don't recommend the right breeds. Often the clerks love animals but don't know the difference between a Husky and a Malamute. I run out of fingers and toes to count the number of times I've overheard pet-shop clerks unintentionally misrepresent a breed or misinform a customer:

"Yes, ma'am, Afghan Hounds are very friendly."

"No, sir, Finnish Spitz don't bark much."

"Yes, ma'am, Chihuahuas love children."

"No, sir, Gordon Setters don't grow very large."

"Yes, ma'am, a Bull Terrier is a cinch to train."

Although pet shops are an excellent source for pet supplies, they should not be *anyone's* choice as a source for a puppy.

How about *individuals* like friends, neighbors, or coworkers? Suppose a colleague says, "I hear you're looking for a Golden Retriever. Our dog Molly recently had pups who are eight weeks old now. Would you like to see them?"

There's always the danger that your personal or work relationship could suffer if you buy a puppy from a colleague and the puppy doesn't work out. But the biggest danger is that your colleague may fall into the "poor breeder" cat-egory we've already shuddered about. Ask him these questions:

1. "What made you breed Molly?"
   Good answer: "We felt the puppies would be of exceptionally good quality."

Steer clear: "To make money . . ." "To teach the kids the miracle of birth . . ." "Molly is just so cute we decided the world needed more cute Golden Retrievers . . ."

2. "Does Molly meet the AKC Golden Retriever Standard?"

   Good answer: "A Golden breeder said she was a little small and her eyes were a little light, but that otherwise she was very nice. He recommended we look for a sire who was the correct size and had dark eyes."

   Steer clear: "Does she meet the *what*?"

3. "Where did you find the sire?"

   Good answer: "Molly's breeder recommended a Golden named Ledgerock's Beemer, from conformation and obedience lines. He was a long drive away, but it was certainly worth it."

   Steer clear: "Oh, he lives right down the street . . ."

4. "Have Molly and Beemer been X-rayed for hip dysplasia and PRA?"

   Good answer: "Oh, yes. We have the certificates."

   Steer clear: "Have they been X-rayed for *what*?"

In an upcoming chapter you'll find other important questions, but if at this point your colleague has given good answers, put him on your list of good breeders.

How about *animal shelters*? Their signs plead: "Please save a life—adopt one of our dogs!"

They've got a point.

Every evening, you drive by the Humane Society on your way home from work. Might there be any Miniature Schnauzers there? It's possible. Many purebreds end up at the animal shelter, and a popular breed may have a representative there, especially at a large city shelter. There are a hundred reasons why any dog, purebred or mixed, may have been dropped off at the animal shelter:

"Our landlord says we can't keep him."

"We're moving to Georgia and he doesn't like hot weather."

"We have a new baby."

"He's not housebroken . . . he chews . . . he barks . . . he digs . . ."

In other words, owners couldn't or wouldn't keep their dogs any longer. Should you adopt one of them?

An important consideration for many prospective buyers is money. Shelter dogs cost $10 to $75, a far cry from $250 to $500.

But what kind of dog will you be getting for ten dollars? Maybe the worst-quality dog in the world. Maybe the best-quality dog in the world. The biggest disadvantage of Humane Societies is the unpredictability of what you'll find inside them.

If you do decide to go in, you'll at least have a big advantage over the person who is impulsively "looking for a dog for the kids." This person will be faced with a bewildering number of decisions, and the barking bedlam of the animal shelter is not the best place to make wise decisions.

You're different. You've already made the decisions. You know you want a purebred and you know what breed you want. In the final step of this book you'll learn how to evaluate a dog's temperament, health, and general structure, and then you'll be ready to apply your knowledge to the dogs at the Humane Society. You'll be able to choose the best one for you.

But you must go in with your eyes open, not just your heart. You don't know the background of these dogs, and thus you're taking a gamble, especially in the important area of health, and especially with those breeds who are susceptible to hip dysplasia and PRA. You should consider the potential heartbreak of dearly loving a dog who is doomed to suffer or die.

On the other hand, the satisfaction and joy that comes from giving such a dog a loving home for a few years may be worth it to you. I can empathize with both sides of this question. I have adopted—and will continue to adopt—both purebred and mixed-breed dogs from our overcrowded humane societies. And I'll continue to buy from good breeders, as well. Good dogs are everywhere, if you only know how to recognize them. So if the option still interests you, call the shelters of nearby large cities and ask them to notify you if a member of the breed you're looking for comes in.

And while you're on the phone, ask them if there is a *breed rescue league* for your breed.

Suppose you're flipping through a magazine and you see a photograph of a pathetic-looking Irish Setter. The ad says: "The Irish Setter Rescue League is desperately looking for good homes for abused and neglected Setters. Please call." Should you call?

If you're looking for an older puppy or adult, and if you have a deep desire to make a difference in the life of a particularly needy dog, do call. But be prepared to answer as many questions as if you were trying to adopt a child. Rescue leagues don't want to save a dog from one bad home simply to place him in another, and so they'll consider you very carefully before handing over a dog, especially potentially aggressive breeds like American Staffordshires or Akitas.

But don't let the "potentially aggressive" label scare you away. Most rescue leagues scrupulously examine their dogs for temperament problems; they put biters to sleep humanely. Thus the dogs that the leagues offer for adoption are often among the gentlest, kindest representatives to be found anywhere. It is both astounding and shaming that many dogs will accept years of neglect without ever losing their kindly dispositions or their faith in mankind.

It's true that some rescue dogs are timid, distrustful, and in need of gentle, careful handling. It's true that some of them have serious behavior problems that will need to be worked out. But many of them have absolutely no behavior problems and are among God's sweetest creatures.

Only *you* know if the idea of taking a hunted, haunted dog and turning his life around appeals to you. If it does, give the rescue leagues a call. Contact PROJECT BREED, Shirley Webber, 18707 Curry Powder Lane, Germantown, MD 20874, (301) 428-3675 to obtain her comprehensive list of rescue organizations.

Finding Breeders

You've decided not to buy from a poor breeder or a pet shop. You've questioned any friends or colleagues with litters. You've made your decision on whether to check out Humane Societies and rescue leagues. Now you'll add some good breeders to your list.

The following nine sources should produce your breeder list. If your breed is popular, you won't have to check out every source, but if it's uncommon, you may have to.

1. National clubs Turn to the Appendix for the address or telephone number of the national club for your breed. Call or write, requesting the names, addresses, and phone numbers of breeders and regional breed clubs in your area.

If you write instead of call, enclose a business-sized self-addressed, stamped envelope. Some clubs will respond within a few weeks; unfortunately, some will take several months. The club may also send you an informational brochure; if it contains photos and an honest assessment of the breed's positive and negative features, fine, but most brochures merely recite the AKC Standard and extol the breed as, "smart, brave, loyal, an ideal pet . . ." They seldom tell you if the breed is restless, nervous, aggressive with other dogs, noisy, and so on.

The biggest advantage of buying from a breeder who belongs to a national or regional club is that such clubs usually maintain codes of ethics that member breeders must sign. This is one indication that the breeders are reputable.

2. AKC Gazette *and UKC* Bloodlines If your breed is an AKC breed, check your library for the AKC's official publication, *Pure-Bred Dogs/American Kennel Gazette.* If your breed is a UKC breed, write or call the United Kennel Club, 100 East Kilgore Rd., Kalamazoo, MI 49001, (616) 343-9020, for a copy of their bimonthly publication, *Bloodlines.* In either magazine, if you find any advertisements for your breed, photocopy the page. Most breeders who advertise here are reputable.

3. *Telephone directories*   While you're at the library, check the yellow pages of the cities in your area. Look under Dogs, Breeders, and Kennels. Photocopy anything on your breed.

4. *Newspaper classified*   Still at the library, photocopy the Pet classified column of local newspapers. The majority of newspaper classifieds belong to Mrs. Dipps, Breeder Biffs, the Greedybuckses, and Snowsnob Kennels, but good breeders occasionally advertise here.

5. *Dog World* and *Dog Fancy* magazines   In these monthly newsstand magazines you'll find many display and classified ads.

6. *Quaker Oats*   I'll bet this source surprises you! Call Quaker Oats in Chicago, Illinois, and request the Gaines booklet entitled *Where to Buy, Board, or Train a Dog*. It lists many breeders by city in each state. Some of these breeders are good breeders, but unfortunately, many of them are poor breeders.

7. *Off-lead* and *Front & Finish magazines*   Breeders who are interested in their dogs' trainability may advertise in these two obedience magazines. The addresses can usually be found in the classified section of *Dog World*, under "Books-Magazines." Order sample copies.

8. *Hunting/Schutzhund magazines*   If your breed is a sporting dog or hunting dog and you're interested in hunting, check the classifieds of outdoor magazines on your newsstand. If your breed is a protective breed and you're interested in the sport of Schutzhund, order a sample copy

of *Dog Sports* magazine; the address can usually be found in a display ad in *Dog World*.

9. *Dog shows*   Want to find hundreds of breeders in one place? Go to a dog show! This may be especially useful if you're still wavering between breeds.

First you have to find a show in your area. Look for a list of upcoming AKC/UKC shows in *Dog World* and *Dog Fancy*, or request a list of rare-breed shows from your rare breed's national club.

For AKC shows, try to arrive by 9:00 A.M. Ask for the booth selling show catalogs. In the front of the booklet you'll find an index of breeds or judging schedule. Look up your breed and you'll see listed the number of representatives at the show, ring numbers, and judging times. Also, head over to the obedience rings and see if your breed is competing in obedience.

If you see someone with your breed, stop and chat, but keep in mind that you're only getting one person's opinion. If someone tells you their Saluki (a reserved breed) loves people, recognize that they have a very unusual Saluki and don't run out and buy one hoping for the same unusual trait in yours! There will always be exceptions to every rule, but you shouldn't depend on single experiences, either positive or negative, when choosing a breed. We've based the profiles in this book on the experiences of hundreds of reputable breeders and owners.

So use your day at the dog show to observe the breeds up close, to chat with breeders and owners, and perhaps even to get a few phone numbers. The names and addresses of all exhibitors are included in the catalog; keep it for future reference. Most breeders won't pressure you into buying a dog, but just in case, don't get carried away by a breeder's glowing descriptions of his dogs or his breed. Don't let any breeder talk you into a breed that you know from your research is not right for you. Be affable, but a hard sell. You're not ready to buy your puppy just yet.

As you're collecting your list of breeders from various sources, you'll be reading through display ads and classified ads. You'll start coming across terms you don't understand, so let's interpret some of that unfamiliar jargon.

You may notice *titles*:

Titles, Part 1   *By Ch. Ravenwood Lad x Ch. McKenna's Sassy Frassy.* These are the parents. The sire's name is after the word "by," and the dam's name is after the word "x" (meaning "out of"). So the sire is Lad, from Ravenwood Kennels, and the dam is Sassy Frassy, from McKenna Kennels or owned by the McKennas. "Ch." means that a dog has won his AKC bench championship: In several dog shows, judges felt that he fit the AKC Standard for his breed better than the other dogs present in the show ring.

Ah, that must mean they're of good quality, then. Well, remember that quality involves health and temperament, too, and although these *should* be taken into consideration at a dog show, that's not always the case. But a bench championship is a good start in gauging quality— if the competition (the other dogs present in the show ring) were good representatives of the breed. If Lad was simply the best of a poor lot, he won what is termed "a cheap championship." Informed buyers don't gasp in awe simply because a dog has a Ch. in front of his name.

Titles, Part 2   *By Dual Ch. Knollwood Jason WC ex. Fld Ch. Sly's Happy Emma WDX.* These are AKC hunting titles, where a dog must do something other than trot around the show ring

and be compared to the standard. He must locate hiding birds, retrieve fallen birds, or track or chase game. Jason is a dual champion: He's won both a bench championship *and* a field championship. Emma has won a field championship (sometimes abbreviated FCH).

Other titles earned by a dog who has demonstrated his hunting or retrieving instincts include WC (Working Certificate), WCX (Working Certificate Excellent), WD (Working Dog), WDX (Working Dog Excellent), JH (Junior Hunter), SH (Senior Hunter), LCM (Lure Courser of Merit), and CG (Certificate of Gameness). If your breed is a sporting, hound, or terrier breed, hunting titles are a solid indication that the breeder is working with his dogs and breeding for working qualities.

Titles, Part 3   *By TJ's High Plains Drifter CD TD ex. OTCh. Aleesha of Pinecrest UD.* These are AKC obedience titles. A CD (Companion Dog) means that the dog obeys basic commands on and off leash. A CDX (Companion Dog Excellent) means that he also retrieves and jumps hurdles. A UD (Utility Dog) means that he also obeys hand signals and locates objects by scent. A TD (Tracking Dog) means that he has demonstrated his ability to follow human scent. An OTCh. (Obedience Trial Championship) means that he has won high awards in many obedience competitions. No matter what the breed, when I have a choice between an ad that mentions obedience titles and one that does not, I always call the ad with the obedience titles.

Titles, Part 4   *By VA Canto vom Schonau SchH III AD FH.* These are German titles; this dog has probably been imported from Germany. The Germans rate the quality of their dogs by letters: VA is Excellent Select and V is Excellent, a

shade lower. AD is a physical endurance test and FH is a tracking test. But the most common German title you'll see is SchH (Schutzhund) and then the Roman numerals I, II, or III, indicating the level of proficiency at this rigorous German sport.

Are German titles something to seek out or something to avoid? Those interested in Schutzhund need a dog bred for competitiveness, but those looking for a family pet might find such a dog too tough, too energetic, or too independent. I'd advise the novice to avoid breeder ads that emphasize German imports and Schutzhund scores.

Titles, Part 5   Other titles include:

HC: Herding Certificate. Border Collies and other herding breeds have demonstrated their instinct to herd sheep, goats, or ducks.

ROM: Register of Merit. Awarded to sires and dams who have produced several champion offspring.

TT: Temperament Tested. A judge has tested an adult dog's reactions to strangers, odd sights, and sudden sounds, and has declared the dog to be of sound, normal temperament.

Besides titles, you'll notice *grades* of puppies:

Grades, Part 1   *Show-quality puppies.* Show-quality puppies are those who, at a still-tender age, show promise of developing all of the characteristics called for in the AKC Standard, even the ones that don't seem that important, for instance, feet shaped exactly so.

Predictions of show quality are risky because the puppy still has to grow up. The younger he is when the show-quality label is hung on him, the more time he has to develop faults that will

hamper his chances in the ring. People who are looking seriously for a show winner should buy a puppy over six months of age. An older show prospect will cost more, of course, because the breeder has spent time and money raising him and now feels more certain about the dog's winning potential.

Breeders are reluctant to sell show prospects to novices because novices have a reputation for changing their minds about showing the dog. Such a reversal is disappointing to the breeder. Some breeders will sell show prospects to novices only on *co-ownership*, where both you and the breeder legally own the dog. This gives the breeder some control in the dog's show and breeding career and gives him the right to use the dog at stud. Co-ownerships are complicated contractual relationships and not recommended for the novice.

Are breeders of show puppies always good breeders? Unfortunately, no. Some show breeders have only one interest in mind: winning championships. Some show breeders look down their noses at the mere mention of working ability. Their dogs may be beautiful, but as we discussed earlier, breeding for any single factor often results in neglecting the other factors, and neglecting a breed's temperament and working ability can change the breed from what it was intended to be.

In addition, genetic diseases are becoming increasingly common in many breeds, and much of the blame goes to show breeders who ignore the fact that their best stud is a proven carrier of hip dysplasia or PRA. More blame should go to the AKC for granting show championships to such carriers, since championships encourage more people to breed to the carrier and foist more unhealthy puppies onto the unsuspecting public. But the majority of the blame for the increase in genetic disease goes to poor breeders and puppy mills, who seldom check their dogs' health before breeding.

Grades, Part 2   *Working puppies.* Here we have the opposite of the show breeder. The working breeder wants to sell his puppies to people who will work them in obedience competition, hunting, herding, Schutzhund, and other activities.

Most working breeders have their breeds' best interests at heart because they're trying to ensure that the breed retains its working instincts. But just as some show breeders have a disdain for working ability, some working breeders have a disdain for the AKC Standard. Their dogs work well but look odd, and neglecting a breed's unique appearance also can change the breed.

I'm sure you realize now that it's the *combination* of appearance, temperament, health, and working instinct that makes each breed special and unique!

Grades, Part 3   *Pet (companion) puppies.* Pet puppies fail to meet the AKC Standard in some way that would hamper their chances for success in the show ring: ears too small or too large, head too narrow or too wide, too many patches of color, and so on.

Ah ha! you say. Finally, we've come to the grade of puppy for me. All this talk about show dogs and working dogs has been interesting, but I only want a pet.

No, you don't—you want a *quality* pet. I'm sure that by now you can recite what quality means: good temperament, good health, and a decent representative of the breed. Don't accept less just because a breeder is calling it "a pet puppy." Don't let a breeder show you a shy, sickly, odd-looking puppy and say, "Oh, all our

show and working puppies have been sold; this one's just a pet." Insist that your pet be a *quality* pet.

Do all breeders divide their puppies into show, working, and pet quality? No. Some breeders, perhaps the wisest ones, make no predictions about their puppies' potential; they sell each puppy at the same price to the best possible home and let nature and fate take its course. When you go to a reliable breeder and choose from a well-bred litter where the puppies seem much alike (called a "uniform litter"), you often stand as good a chance as anybody of getting a puppy who could turn out to be a show winner or working competitor as well as an outstanding pet.

Besides titles and grades, you'll notice other terms:

Whelped 6-25: The litter was born June 25.

Imported stock: The breeder has imported his dogs, usually from the breed's country of origin. This claim is only as good or bad as the individual quality of each imported dog, but usually the breeder hasn't gone to all that trouble and expense to import poor dogs.

Pointed stock: The breeder's dogs have attained some points toward their AKC championships. Points are accumulated by being judged better than other dogs. Fifteen points makes a champion.

Litter is linebred on Ch. Shazaam: The sire and/or dam of the litter is closely related to Ch. Shazaam. It doesn't mean much unless you've seen Ch. Shazaam and are satisfied that he is a quality dog.

Reservations being accepted for spring litter: Many breeders take reservations for their puppies before they undertake the breeding. Some of the breeders you will soon be speaking to may have a current litter, but the puppies are already spoken for. For a future litter you'll have to put down a deposit for a puppy and the breeder will notify you when the litter is born.

Home-raised and/or *home-socialized*: Puppies raised indoors grow accustomed to the sights and sounds of a normal home, which is a distinct advantage when the puppy goes to a new home. "Socialized" puppies have been carefully introduced to all sorts of people, sights, sounds, and situations, so that the puppy has learned to accept new things with calmness and confidence.

I once visited a home where four German Shepherd puppies sat behind a baby's tension gate off the living room. When a toy-laden toddler rushed into the room and tripped, toys crashed with a huge clatter! All four puppies simply perked their ears with interest and wagged their tails at the strewn toys, the bawling toddler, and the consoling mother. I chose "Luke," my dear companion and enthusiastic show/obedience/Schutzhund competitor, from this beautifully home-raised and home-socialized litter.

Temperament Tested: The breeder has tested each puppy to determine its personality and the type of family it would best fit into. Breeders who take the time to test temperaments care very much about their puppies' futures. You'll soon be learning how to do your own temperament testing.

Hips X-rayed, OFA, eyes cleared, CERF: If your breed is prone to genetic disorders like hip dysplasia or PRA, look for these health assurances.

Guaranteed for hips, eyes, health, and/or temperament: We'll discuss later the specifics of good guarantees, but any offer of a guarantee is nice to see in an ad.

Prices: $250 to $500, perhaps more for some breeds, show prospects, older puppies, and adults. If a breeder tries to rope you into more, keep looking. Very likely you'll find a good breeder with more reasonable prices.

Now let's combine this new terminology into three sample ads. What would a good ad say?

"Labrador pups for show, field, obedience, or pet, whelped 4-26. By Powderhorn Dusty Tamarack CD ex. Glenmoor's Abbey TT (pointed). Both parents OFA and CERF, occasionally hunted. Home-raised, socialized, temperament-tested. Guaranteed. $375. Wildwood Kennels." This ad says all the right things. The breeder is interested in all aspects of the Labrador—the sire works in obedience, the dam has a temperament certificate and some conformation points, both parents have been checked for hip dysplasia and PRA, and both parents have been exposed to hunting. The puppies have been prepared for a home environment, they are guaranteed, and the price is reasonable. Most importantly, the tone and terminology of the ad sounds as though the breeder knows what he's talking about.

You'll still have lots of questions for this breeder when you call him, but he is a breeder to include on your list.

In contrast, let's look at an ad that has ignorance written all over it: "AKC purebred Dobies for sale! Champion pedigreed! Giant-sized! Rare colors! Shots and wormed! Males $90, females $75!" I saw this ad in a newspaper. What's wrong with it?

1. If the puppies are AKC registered, they have to be purebred with a pedigree, don't they? ("Pedigreed" is an incorrect term, and its use should always warn you that you are

dealing with an unknowledgeable breeder.)

2. A knowledgeable breeder would never use the affectionate nickname "Dobies" in an advertisement.

3. Dobermans are not supposed to be "giant-sized." One of the breed's attributes is its medium-sized athletic build. Any breeder who decides to change the AKC Standard by making Dobermans as big as Great Danes doesn't care much about preserving the breed as it was intended to be. If this breeder wants to create a new breed, let him do it without changing the magnificent Doberman.

4. Dobermans come in black, red, blue, or fawn. Although blue and fawn are not as common as black and red, they are not "rare." But maybe the dogs are white—some breeders deliberately breed white Dobermans, which is indeed a rare color because it is a disqualification. Good breeders respect the AKC Doberman Standard enough not to deliberately breed dogs with disqualifications.

5. Since all puppies should be given initial inoculations and checked for worms, good breeders seldom include this assurance in their ads. However, poor breeders have so little else they can say about the litter that they might push "shots and wormed" as the big selling point.

6. The difference in price based on the sex of the puppy is often indicative of a poor breeder. Good breeders base any difference in price on the quality (or grade) of each puppy, regardless of its sex.

7. The worst thing about this ad is its overall tone. The incorrect terminology and exclamation point extravaganza reveal it as the ad of a poor breeder.

Finally, let's look at perhaps the most common newspaper ad. "English Setter puppies. AKC-registered, raised in our home, happy, friendly, healthy. $175." This ad doesn't say much, but there's nothing really wrong with it, so you can put this breeder on your list to call, especially if you're short on funds and haven't been able to find many English Setter breeders in your area.

What if you haven't found *any* breeders in your area? What if your breed is so uncommon that the nearest breeder is six states away. His ad says, "We will ship." He will select a puppy for you, secure him in a shipping crate, put him on a plane, and you'll meet him at your end of the flight. Is this a good idea?

In the case of a dog over four months of age, probably not. The temperament of an older dog is more settled than that of a young puppy, and you should take an up-close, personal look at him before committing yourself. Although the breeder may be willing to ship a dog on a trial basis, after which you can ship him back if you and he don't hit it off, this is a lot of trouble and expense for you and an unnerving experience for the dog.

In the case of a young puppy, it's fine if the breeder is extremely reputable. You should speak extensively and earnestly with him on the telephone, because you must be sure you can trust him to select exactly the puppy you're looking for. With uncommon breeds you may not have another choice. And sometimes you don't even have this choice—many breeders refuse to ship puppies out of concern that something might go wrong. This concern is justified, because tragedies can and do occur when animals are shipped.

So, if possible, try to find breeders who live nearby. However, if every breeder in your state is asking a ridiculous price, consider contacting a breeder in another state and having a puppy shipped. As we've said, with a young puppy from a reputable breeder who stands behind his dogs with health checks and health guarantees, this is a perfectly viable alternative. You'll have to pay for shipping, but you can gain satisfaction in knowing that you're not padding the wallet of a greedy breeder and that you're helping discourage future high prices. If *every* breeder in *every* state is charging a ridiculous price, seriously consider choosing a different breed.

8

Contacting Breeders

You've assembled your list of breeders from a variety of sources. You've used your knowledge of correct and incorrect terminology to weed out the obviously bad sources. You're ready to start calling the most promising ones. But you're nervous because you don't know what to say.

Let's listen in on a telephone conversation between a prospective buyer and a good German Shepherd breeder who asks all the right questions at the right times. You'll seldom be lucky enough to get one of these perfect breeders, but you'll learn from this conversation the important topics of discussion. Then, if the breeder doesn't bring them up, you'll be able to.

If you're looking for a puppy under four months old, you can use this conversation almost word for word. If you're looking for an older dog, read through the conversation for general information, then read the section following for specific advice about the older dog.

When you do call the breeder, you'll want to have a pen and a pad of paper handy so you can take notes.

YOU: Mr. Kelly, my name is Kim Vickers. I'm interested in German Shepherds and I was given your name by the German Shepherd Dog Club of America. Might you have any puppies available?

If no, "Do you know of another breeder in the area who might have a litter?" If no, "Do you have any litters planned in the near future?" If yes, tell him you'll get back to him if you can't find any current litters. If you do get back to him, run through as much of the conversation as possible, then ask to visit him to decide if his litter would be worth putting a deposit on and waiting.

If you're lucky and the breeder does have a litter:

BREEDER: Yes, I do have two puppies from my current litter.

YOU: Do both parents have OFA hips?

This should be the *first* question you ask of any breeder who has a puppy available. If your breed profile says that the parents should be OFA-certified for hip dysplasia, CERF-certified

228

for PRA, or VWD-tested for bleeding disorders, *be sure* they are.

Do not buy a puppy whose parents have not been X-rayed. Do not accept these excuses: "Well, um, no, but we've never had a problem with it. . . . X-rays don't prove anything, because that disease isn't really genetic. . . . We're going to get them X-rayed soon. . . . We didn't want to spend the money. . . . We don't X-ray, but we guarantee all our puppies and if he develops any problems, just bring him on back and we'll give you a new one."

You should reply, "I'm sorry, but I'm only buying a puppy from a breeder who X-rays. Thank you for your time."

Stand firm. It's heartbreaking to discover, months down the road, that your beloved puppy is showing the first signs of a crippling disease that will cause him pain and suffering and perhaps cost him his vision or his life. It probably could have been prevented if the breeder had been more careful.

Note: Some breeders do not send their X-rays to the OFA and CERF, but they do X-ray their dogs. If the X-rays are accompanied by a letter from a veterinarian attesting that the dogs are free of hip dysplasia or PRA or cataracts, this is usually fine.

BREEDER: Yes, I have their OFA certificates on hand. What kind of puppy were you looking for?

Your answer will be one of these three:

One who can *work* in obedience competition, herding, hunting, etc. Caution: If you want a working dog, buy only from a breeder whose dogs have *proven* they can work. Not all English Cockers, for example, can hunt. Can the breeder's dogs put on a demonstration of their working ability? If not, thank the breeder politely and go on to your next name.

One to *show* in the conformation ring. It's difficult to determine if a puppy will grow up to have the specific AKC Standard traits needed to win in the show ring. As a novice, you'll have to leave the choice almost entirely up to the breeder, thus it's crucial to find a breeder you can trust. If you're seriously looking for a winner, dig into your wallet and buy an older dog.

Since breeders are reluctant to sell show prospects to novices, you should already have done some research. Then when the breeder asks about your experience, you can tell him that you haven't shown yet, but that you've gone to a show, talked to breeders and exhibitors, and read the latest book on Pomeranians. You can tell him that you're planning to join your local breed club and attend handling classes with the puppy. These words and efforts might reassure him.

A *pet*. A family pet, a child's playmate, a jogging companion, a lapdog.

YOU: I'm looking for a good companion. My husband and I are very active, and we'd like a dog to go hiking with us and to play with our two children. How old are the puppies?

Seven to twelve weeks is the optimum age to bring a puppy home, but an older dog will probably settle just as nicely into your home if he's been raised in a home and is familiar with family life. Avoid older dogs who have been raised in a kennel run; they may have "kennel syndrome," a deeply imprinted fear of new people, sights, sounds, and situations.

Advantages of a young puppy: He comes to you as a clean slate with no bad habits. You raise him and train him right from scratch. You can watch him grow up.

Disadvantages of a young puppy: He requires much care and supervision, and several weeks or months of housebreaking. You can't be absolutely sure of how his personality and habits might turn out.

Advantages of an older puppy: He's not as

delicate as a young puppy and doesn't require as much adherence to a strict schedule. He's more suited to active play and exercise. He may already have had some housebreaking and training. You can better determine what his appearance and personality will be like.

Disadvantages of an older puppy: An adolescent puppy, like an adolescent child, may be extremely active and mischievous as he goes through physical and emotional stages of awkwardness, flightiness, and rebellion. Dominance testing may raise its ugly head, and since the two of you haven't yet established a pecking order, you might be in for some frustrating leadership tests. You have to be both very patient and very firm as you nip bad habits in the bud and establish some control.

Advantages of an adult: His appearance and personality are fairly settled. He's probably housebroken and may have had some training. You've bypassed both his delicate baby stage and his frustrating adolescent stage.

Disadvantages of an adult: An adult is not as pliable about learning new things, so his bad habits may be difficult to change. His adult self-confidence may make him resistant to new leadership and new rules. You missed his fun growing-up years, and you'll never know exactly what experiences he's had or what things he's seen.

BREEDER: They're nine weeks old.

YOU: Males or females?

Males, called dogs, versus females, called bitches—which makes the best pet? The debate rages.

Very generally, males tend to be more independent, more apt to test their owners, more stable and reliable in mood. If they're aggressive, they're aggressive most of the time, and if they're easygoing, they're easygoing most of

the time. They can fight, so if there are male dogs running loose in your neighborhood, perhaps you shouldn't get a male. They also lift their legs, so if you have lots of expensive shrubbery, keep him away from it or else don't get a male.

Very generally, females tend to be quicker to learn, more submissive, and more moody. They can be sweet one day and grumpy the next. They can be sulky about correction. They're easier to walk than males because they don't try to lift their legs on every bush, hydrant, and telephone pole.

In certain breeds or certain breeding lines, these male/female generalizations may be reversed, so ask the breeder for his opinion on male and female personalities in his breed.

Let's touch on male and female reproduction. Unless you're buying a top-quality dog for showing and breeding, don't plan on recouping your investment by selling your male's stud services or your female's puppies. It is unlikely that your pet puppy will develop his promise of quality to the point that he is worthy of passing his genes on. Just love him for what he is and for what he turns out to be and don't put conditions on your love that can lead to disappointment.

If you don't neuter your female, she will come into season or "heat" twice a year. This is a three-week affair. It may be messy or she may lick herself clean; she may act restless, silly, nervous, or perfectly normal. She should never be allowed off the leash during her heats unless you live in the country and have high fences that are planted in the ground to prevent digging under them. If you also own an unneutered male, he will be a brat during your female's heat period; he'll whine and pace and bother her mercilessly. Heat periods are no fun at all.

If you don't neuter your male, you may discover to your dismay that your adolescent puppy is a sex maniac—he sniffs eagerly at female dogs (even spayed ones), climbs all over children (sometimes growling at them and frightening them), fights with potential rivals, and ignores your commands because his eyeballs are busy roving up and down the street. Any male who exhibits these traits should be neutered. No, breeding him *won't* help—it will make things worse! Males who have been bred may try to lift their legs in the house as a sign of dominance and masculinity. Just what you need, right?

I highly recommend neutering *all* pet females and any pet male who is highly sexed, energetic, restless, or aggressive. Neutering prevents serious reproductive cancers and often calms excitable dogs. Neutering does not make a dog fat—overfeeding makes a dog fat. If your neutered dog is fed slightly less food than before (since his metabolism no longer requires so much), he will stay slim and trim.

Now let's return to our imaginary telephone conversation. The breeder has only females.

BREEDER: They're both females, but they're not for show or breeding. I'm selling them as pets with a spay contract.

Oh-oh, what's a spay contract? It's a sign of a good breeder. He feels that his two females do not meet the German Shepherd Standard to his satisfaction, and he wants them spayed so that they cannot be bred. Breeders who use spay contracts have the breed's best interests at heart.

With a spay contract, the puppy's papers are shown to you, but they're withheld until you send the breeder a copy of the veterinary spay certificate. It's also possible, but not as common, for breeders to require a male puppy to be neutered.

Before you buy a puppy with a spay contract, ask about the faults that are causing the breeder to sell the puppy as a pet. Light-colored eyes or cowhocks or a head that is too narrow certainly will not affect the puppy's ability to be a wonderful pet. But if the fault is in the puppy's coat, ears, or overall color, keep in mind that these are the areas where even people who don't know much about dogs notice whether a dog looks like his breed. To own an appropriate-looking dog, you should steer clear of such noticeable faults and disqualifications as long coat in a shorthaired breed, hanging ears in a prick-ear breed, patches in a solid-colored breed, and so on.

Two other faults that cause a breeder to sell a puppy as a pet are *monorchidism* and *cryptorchidism*. A male puppy with only one testicle descended is a monorchid; he's capable of breeding but shouldn't be allowed to because the condition is hereditary. A male puppy with neither testicle descended is a cryptorchid; he's incapable of breeding. A monorchid or cryptorchid makes a fine pet, but to prevent cancer problems with the undescended testicle(s), the dog should be neutered. Unfortunately, neutering can be tricky because the veterinarian must dig deep to reach the undescended testicle(s).

What are the faults of our imaginary breeder's puppies?

YOU: I'm not looking for a puppy for show or breeding, but what faults do they have that are making you sell them as pets with a spay contract?

BREEDER: Nothing major. They're short in the body, they don't have enough angulation, and they have a high tail-set.

Your brow is furrowing, but don't worry. These details are important in the show ring but will never be noticed by you or anyone else.

YOU: What color are they?

Listen for correct terminology. If your breed profile describes a color as "sable," the breeder should not say, "Oh, sort of a mixed brownish blackish."

BREEDER: One is black-and-tan, one is sable.

YOU: What's the breeding on the puppies? What are their lines?

This question is another key one, because it helps separate the wheat from the chaff. You are using canine terminology to ask about the puppies' pedigree and background. Knowledgeable breeders will understand the question and discuss the sire, dam, perhaps a noteworthy relative, any titles his dogs have won, any special training they have. Unknowledgeable breeders won't understand the question or they'll say something funny, like "Oh, the parents are both Basset Hounds and they even have AKC papers! These are champion-pedigreed puppies!"

BREEDER: The sire is a champion who's line-bred on Champion Covy-Tucker Hill's Manhattan. I have pictures of him and I also have pictures of some of his offspring so you can see the type of puppy he produces. The dam is champion-sired and has her CD. This is her second litter and I have pictures of her other litter. She has mixed American and German lines.

I'll bet you're surprised at how much of that you actually understood! The sire has his bench championship; "Manhattan" is a show dog in his background. The dam has a champion father and also a Companion Dog obedience title.

The breeder seldom owns the sire because a good breeder searches for the stud dog that will best complement his bitch. If he simply used whatever male happened to be handy, you might suspect him of cutting corners elsewhere as well. Of course, it may be that he used one of his own studs because the dog was perfect for the dam.

YOU: Could you tell me a little about their temperaments?

You want more than, "Oh, they have good temperaments." That means different things to different people. Are they energetic, boisterous, high-spirited? Calm, easygoing, gentle, quiet, laid-back? Tough, aggressive, independent? Nervous or skittish? With strangers, are they friendly, aloof, aggressive, or timid? If "protective," can people go near them? Ask for details.

BREEDER: The sire is confident and easygoing and doesn't get excited about anything. The dam is alert and she'll bark at you and watch you, but she wouldn't bite you. She loves to work in obedience. I'd be happy to show you how well she works.

YOU: Have the puppies been raised indoors? Have they been socialized?

Another terminology test: Good breeders will understand what "socialized" means; poor breeders will not.

BREEDER: They spend most of their time in their outdoor yard, which has a swinging door into a heated kennel. But we do bring them indoors and we've spent a lot of time playing with them and introducing them to people. They're outgoing puppies, very happy and sure of themselves. By the way, how did you become interested in German Shepherds? Have you owned them before?

The breeder is being cautious, trying to find out if you want the breed just because you watched an old episode of "Rin Tin Tin" last night. He may ask questions about your children, other pets, living environment, and working hours. Don't be offended by his concern.

YOU: My husband and I owned mixed breeds when we were growing up, but when we decided to get a dog for our family, we did some research and decided that a German Shepherd was the breed we were looking for. We like their intelligence, their reserve, and their devotion to their family. We do understand that they need exercise and obedience training, and we're looking forward to that. We also have a fenced yard.

BREEDER: You have a good attitude. I think you'll find these puppies to be just what you're looking for.

YOU: What price are you asking?

BREEDER: Three hundred and fifty dollars.

YOU: I'd like to see them. When would be a good time?

Now if you're interested in an older dog, open your telephone conversation with this:

YOU: I wondered if you might have any older puppies (adults) available as pets. I'm looking for a dog who is housebroken (has had some obedience training, will be reliable with my eight-year-old, will get along with our pet rabbit, and so on).

If the breeder has a suitable prospect, ask many of the same questions you would for a young puppy: appropriate health certificates, age, sex, color, parental background, temperament and personality, whether it's a house dog or kennel dog, and price.

In addition, the following questions will help you decide whether he warrants the trip to see him:

1. How long have you had him? Why are you selling him?
2. I'll be bringing my kids with me—how will he get along with them?
3. What type of training has he had? House-breaking? Any obedience? Was he easy or difficult to train? How does he walk on a leash? How does he ride in a car?
4. How is he with other dogs? Cats?
5. Is he a barker? A digger? A chewer? If there is some character trait or behavior trait that you and your family cannot tolerate, make it clear to the breeder.

If you've finished listening to our imaginary phone conversation but you still don't have the nerve to make your own call, write a letter. Breeders prefer phone calls so that they can get a feel for the type of person you are, but if the telephone makes you unbearably nervous and if your breeders are located some distance away, it's okay to write letters initially and then call those whose responses you liked best.

Dear Mr. Kelly,

My husband, children, and I are looking for a German Shepherd puppy for a family companion. We've done some research and feel that a German Shepherd is suited to us. We like their intelligence and devotion. We know that they need plenty of exercise and training, and that suits us perfectly. We're outdoor enthusiasts who like to hike. We have a fenced yard.

We've both enjoyed mixed-breed dogs when we were growing up, so we have some experience with dogs. We're looking for a male puppy, preferably black-and-tan, seven to twelve weeks old. The sire and dam must be OFA-certified and have excellent temperaments.

Might you have a puppy available? Could you tell us about your breeding program? We'll look forward to hearing from you.

Enclose a self-addressed stamped envelope. When you get some replies, swallow your nervousness and telephone the breeder whose letter gives you the impression that he'll be most friendly and helpful. Assuming that his answers to your questions are satisfactory, you'll head out to his home!

(Bring cash—most breeders are reluctant to take a check.)

At the Breeder's Home

Don't be surprised if the kennel is one in name only. Many breeders own only a few dogs, most of whom may live in the house, and have only a couple of outside runs or a large yard for exercise. Rare today are large kennels with runs for twenty dogs, few of whom ever see the breeder's home. If a kennel is huge and offers you a half-dozen breeds to choose from, thank the breeder, climb back into your car, and leave. These breeders are spreading themselves far too thin and probably know very little about each breed.

However simple or elaborate, any area accessible to the dogs *must* be clean and safe. The yard should be cleaned up, water bowls should be fresh and full, food bowls should not be overrun by ants, and doghouses and fences should be sturdy, with no sharp edges or jagged holes or protruding nails. There should be no broken glass or tin cans littering the ground.

Clean and *safe* are the operative words. Trust your instincts, and drive away if the place looks dirty or dangerous. A breeder who doesn't care about his dogs' safety is not the breeder for you.

What can the adult dogs at the breeder's home tell you? A great deal.

Health: Adults should be clean and brushed. Their coats shouldn't be matted or filthy or have patches of missing hair. They shouldn't be limping or scratching, unless a dog is simply nursing a sprained leg or battling a skin condition. Ask.

Temperament: Adults may rush up to greet you, they may bark furiously at you, or they may stand back and look you over. All of these responses to a stranger's approach are normal, and the response you'll see will depend upon the breed you're looking at. Salukis won't jump all over you, but Bichon Frises might. However, no matter what the breed, the adults should never try to bite, never spook if you happen to shift your feet, and never cringe away from your hand with tails tucked between their legs.

Most dogs are bolder behind a fence, where

even sweet dogs may pretend to be maniacal killers! If you find the behavior of any dog disconcerting, ask the breeder if the dog is related to the puppies you came to see. If so, ask to see the dog outside the fence. An individual dog may be a bit timid or aggressive, but if several of the adults act strange, this breeding line is not the one you want to be involved with.

The dam of the litter is an exception because she may display an understandably suspicious instinct to protect her puppies. Ask to see her away from the puppies so you can more fairly judge her temperament.

Appearance: Since you'd like to own a puppy who will grow up to resemble his breed reasonably well, you'll use your breed profile to pick out serious faults in the adults.

You may be asking, "Why do I have to spend all this time evaluating the adults when I just want to buy a puppy?" Canine authority Roger Caras wisely advises, "Never buy the puppy. Buy the dog the puppy one day will be." The puppies you'll see today, no matter how cute, are likely going to grow up into something very similar to the adult dogs you see. The way you raise your puppy will also influence his temperament and behavior, but you can't change the genes he was born with. Those fixed genes of a purebred can act against you if you look at a cute puppy and rationalize that surely he'll not turn out like the spooky adults in the breeder's yard. Use those fixed genes to act *for* you by assuring yourself that the adults you see *are* the kind of dog you want your puppy to become.

As the breeder is showing you around, form an impression of him. Is he proud of his dogs? Does he pet them and speak fondly to them? Can he control them? Are they happy to see him?

Ask him how long he has owned the breed, why he chose it, what other breeds he has owned. You don't want a breeder who hops from breed to breed, learning little about each.

Ask him what he likes best and least about the breed. An honest breeder will admit negative points as well as positive.

Ask him how often he has litters; the same female should not be bred twice in one year.

Notice whether he uses correct terminology. One term you hope *not* to hear is "thoroughbred" instead of "purebred." A prospective buyer I know once visited a kennel where the breeder declared proudly, "I used to raise wolf-huskies, but they weren't as fierce as my thoroughbred Dobies!" A Thoroughbred is a breed of horse and has nothing whatsoever to do with dogs. Similarly, "full-blooded" is a term never used by knowledgeable breeders.

Since you wouldn't have come unless the breeder's telephone answers were satisfactory, the responses you'll hear in person should also be satisfactory. But you're making a major addition to your life, so don't be embarrassed about double-checking his expertise as a breeder; *he* isn't embarrassed about checking *you* out as an owner. Both of you are doing the right thing.

Will you have to sign anything?

The breeder may show you some of his paperwork before you even see the puppies. But whether he shows you before or after, be sure everything is in order.

For AKC breeds, the breeder must have available a blue AKC registration application for each puppy. The AKC sent him these blue forms, one for each puppy, when the breeder notified them that he had a new litter. If the breeder says that he notified the AKC and they simply haven't sent him the blue forms yet, be cautious. Before you buy, insist upon obtaining a written statement with the puppy's breed, sex, color, date of birth, registered names and numbers of sire and dam, and his name and signature. Then if the blue forms "never arrive," you can use the written statement to register your puppy.

On the front of the blue form will be:

1. Blank spaces for your choice of your puppy's registered name. You can wait until you get home to think of one.
2. Blank spaces for the breeder to fill in the sex and color of your puppy before you leave.
3. Litter information typed in by the AKC: breed, sire's and dam's registered names and registration numbers, date of birth, litter registration number, and breeder's name and address.

The back of the blue form is divided into two sections. The top section must be filled in by the breeder before you leave: your name, address, signature, and date. The bottom section will be filled in by you when you mail the application to the AKC: your name, address, and signature.

When the AKC receives this application, they'll mail you your puppy's white AKC registration certificate—his "papers."

A word about your puppy's "papers." When you buy a purebred puppy that the breeder says is eligible for registration with the AKC, you are *entitled* to a blue registration application. When you buy an older dog, the breeder may already have filled out the blue registration application and received back the white registration certificate. Since it will be in his name, he must sign the white registration certificate over to you—like signing over a car registration.

The only exception to your receiving either the application or the certificate is if the breeder is selling the puppy with a spay contract. Then he may withhold the application or certificate until you send him the veterinary spay certificate.

Don't buy from any breeder who says, ". . . and for twenty dollars more you can buy his papers" or ". . . and I'll charge you less if you let me keep his papers." There should be no financial deal associated with your puppy's papers; the only money you pay for them is the AKC processing fee when you mail in the application.

Besides the blue AKC registration application or white AKC registration certificate, the breeder should also have a *pedigree* ready for you. The pedigree is the litter's family tree. The breeder may have written down only three generations, or he may have taken the time and trouble to do five or six generations.

Linked with the dogs' names, you may see titles that you're now familiar with, like Ch. (bench champion), CD or CDX or UD (obedience), TD (tracking), FLDCH or FCH or WC or WD (field/hunting), SchH (protection), TT (temperament tested), and so on. These are all worthy accomplishments.

The names on the pedigree won't mean much to you, but notice whether there are many names like Duke XXII, White Princess Dutchess Mandy, King John XI, My Little Susie Pie. These all sound like "call names"—the everyday names we give our pets.

A purebred's registered name is traditionally more formal than his call name. For example, Duke's registered name could have been Eaglecraft's Lightning Bolt, the "Eaglecraft" being his kennel, and the "Lightning Bolt" a catchy, impressive phrase. He'd still be called "Duke," of course, but Eaglecraft's Lightning Bolt would be his registered name.

However, since Duke's owners were unaware of this canine tradition, they registered him by the same name they called him. The AKC had to add "XXII" (the twenty-second) because there were so many other unaware people who didn't know they could give their dogs more individualistic registered names. There's nothing technically wrong with Duke XXII, but it does indicate an ignorance about canine traditions. His owners may then have bred Duke with the same ignorance about proper breeding and proper care of a litter.

This is something to watch for, because the pedigrees of poor-quality puppies and pet-shop puppies are often classically recognizable. They have one Ch. Eaglecraft's Lightning Bolt and the rest of the pedigree is peppered with Duke XXIIs and My Little Susie Pies. (Exception: Border Collies, Australian Shepherds, and Australian Cattle Dogs traditionally *do* have registered names that are the same as their call names, especially if they're from working lines rather than show lines.)

To give the pedigree more meaning, ask the breeder if he has photos of any of the dogs. Ask him to tell you about their temperaments and backgrounds. Knowledgeable breeders enjoy talking about their puppies' ancestry and you can learn about the kind of dog your puppy might turn out to be by letting the breeder add descriptions and personalities to otherwise meaningless names. Most breeders have a shoebox full of pictures and will whip it out on a moment's notice.

The breeder should also have ready a *sales contract*. Stay away from any contracts that stipulate co-ownerships with the breeder or that make any demands that you breed your dog or show him. The contract can stipulate that the dog be kept in a fenced yard, that you may not resell him without notifying the breeder, that

the breeder may "grab him back" if you are neglecting him or abusing him, but otherwise, if you pay good money for a dog and are a good owner, the dog should be yours.

The breeder should have ready a *veterinary health certificate* certifying that the litter has been given first inoculations and checked for worms.

The breeder should have ready a *health guarantee* stating that you can get your money back if your veterinarian declares your puppy unhealthy within forty-eight to seventy-two hours. Some breeders also guarantee that if the puppy develops an unstable temperament or congenital disorder within two years, they will *replace* him; however, except in the case of poor temperament, few owners are willing to give up a dog who has become a member of their family. Some breeders instead promise you an *additional* puppy if you neuter the first one; however, many owners don't want two dogs. The best guarantee is for the breeder to *pay* some of the dog's medical expenses; unfortunately, few breeders offer it, feeling that if they have X-rayed their breeding stock, they have already safeguarded as best they could.

The breeder should have ready the official OFA or CERF or VWD certificates attesting that the sire and dam have been found clear. Make sure you see these certificates, and that the names on them are those of the sire and dam.

The breeder may have photographs of the parents for you. The breeder may have the puppies' feeding schedule written up—the food they're accustomed to eating, how much, and when.

Now let's go look at the puppies you came to see!

STEP FOUR

✦

Choosing the Right Puppy

Evaluating a Litter of Puppies

We're assuming that the breeder is allowing you to evaluate the litter yourself and choose your own puppy. However, if you are a novice and the breeder is very experienced, he (or she) may have a firm policy of making the choice for you. Many reputable breeders take great (and usually justified) pride in their knowledge of their breed, of their particular line, and of their puppies. They have been observing the litter since it was born, they may have already temperament tested the puppies, and they may have formed their opinion as to exactly what type of home and owner each puppy would be best suited to. An experienced breeder is usually far more qualified than yourself to make this decision, and you may choose to either accept their decision or to go elsewhere.

But if the breeder is allowing you to make a choice: there they are! One or two or six or eight cute, fuzzy, wriggling puppies. Your puppy may be among them!

As you approach the litter, try not to let yourself be immediately captivated by the biggest or boldest puppy. Your first look should be at the entire litter as a group.

If there are seven puppies and six of them are running away, I'm sorry to say that your visit is over. No, you shouldn't buy the seventh puppy; the chances are too great that the shyness simply hasn't caught up to him yet.

Let the puppies show you their true temperaments, whatever they happen to be. Don't coax a frightened puppy to be brave, trying to convince yourself that you'll be able to "bring him out of his shell." You don't know what's going on in this puppy's mind or genes. Shy puppies usually become shy dogs who'll snap at anything that startles them—say, a skipping child. Someday you might want to bring home a shy puppy as a rehabilitation project, but wait until you have more experience with dogs.

If the litter isn't running away, what should they be doing? Normal puppies of most breeds are friendly. They'll mill around your feet, climb up your legs, tug at your shoelaces, nibble on your fingers. They'll crawl all over one another and wrestle. Puppies of independent breeds like the Norwegian Elkhound might prefer to wander away and explore than to play with you, but they'll still be completely unafraid of you.

Stand and watch the entire litter for a few minutes. You can tell a lot about the individual puppies by the way they interact with one another. Which pups are strong, outgoing, bossy, noisy? Which are quiet, submissive, gentle? Which pups win and lose the tugs-of-war and tussles? Since most families do best with a puppy who is neither the boss of the litter nor the lowest on the totem pole, look for puppies who do not initiate the games and fights, but who do join in and hold their own.

Now clap your hands gently, snap your fingers, jingle your car keys, shuffle your feet, whistle softly, cluck your tongue. Which pups are alert to the sound? Which pups come over to investigate? Which pups are startled or afraid? Most people will do best with a puppy who is attentive and curious. A nervous puppy will not do well in today's busy world and may grow into a skittish hundred-pound adult who is difficult to drag around. A puppy who is completely oblivious may be too insensitive or independent.

If all of the puppies seem apathetic or drowsy, ask the breeder if they've just been eating or playing, because these activities can slow puppies down. A considerate breeder tries to arrange things so the puppies are alert when the prospective buyer arrives.

Now ask the breeder if you can see each puppy available for sale individually, so that you can do a few simple tests of the puppy's temperament. Reassure the breeder that you will not harm or frighten the puppy in any way, that you're basically going to be petting and playing with the puppy, and that you don't mind his sticking around to keep an eye on things. Ask that the first puppy be brought to you in a different room or in a fenced yard and placed on the floor. The breeder and your family may watch from a short distance so long as they do not play with the puppy or distract him.

As you do each test, you'll label the puppy's response (probably mentally, since you may feel self-conscious about writing things down in front of the breeder) as normal, dominant, submissive, or independent.

TEST ONE

Kneel five feet from the puppy and gently clap your hands.

NORMAL: When he notices you, the puppy comes readily to you, his tail up, his expression happy and bright.

DOMINANT: He rushes wildly at you and jumps or nips.

SUBMISSIVE: He comes hesitantly; his tail may be down, his expression uncertain.

INDEPENDENT: Even after he's noticed you, he doesn't come to you; he may wander off to explore.

TEST TWO

Walk away from the puppy, bending down and gently clapping your hands to encourage him to follow.

NORMAL: He follows readily, tail up.

DOMINANT: He tangles your feet badly or nips your pant legs.

SUBMISSIVE: He follows hesitantly, tail down.

INDEPENDENT: He wanders off to explore.

TEST THREE

Gently roll the puppy onto his back and hold him there, with one hand on his chest, for fifteen seconds.

NORMAL: He struggles a little, then settles down.

DOMINANT/INDEPENDENT: He struggles fiercely the whole time, perhaps nipping or yelping.

SUBMISSIVE: He lies passively, perhaps licking your fingers.

TEST FOUR

Lace your fingers together and cradle them, palms up, under the puppy's tummy. Raise him up so that his four feet are just off the floor. Hold him there for fifteen seconds.

NORMAL: He struggles a little, then settles down.

DOMINANT/INDEPENDENT: He struggles fiercely the whole time, perhaps nipping or yelping.

SUBMISSIVE: He hangs passively, perhaps licking your fingers.

TEST FIVE

Sit beside the puppy and stroke him from head to tail along his back repeatedly and firmly.

NORMAL: He wriggles with pleasure or plays gently with you.

DOMINANT: He jumps wildly on you or nips your fingers.

SUBMISSIVE: He rolls onto his back to have his tummy rubbed.

INDEPENDENT: He doesn't like the petting or wanders away.

TEST SIX

Get the puppy's attention with a ball or toy. Roll it across the floor and encourage the puppy to run after it and bring it back.

IDEAL: He chases it, picks it up, brings it back, and drops it or lets you take it from him.

NORMAL: He chases it, picks it up, carries it away to chew on it, but lets you come over and take it from him.

DOMINANT: He chases it, picks it up, carries it away to chew on it, and growls at you or holds on stubbornly when you come over to take it from him.

SUBMISSIVE: He's nervous when the toy rolls past; he's hesitant to approach the toy when it stops rolling.

INDEPENDENT: He shows little or no interest in chasing or picking up the toy.

When you've done the six tests with the first puppy, add up his normals, dominants, submissives, and independents. Most families, especially those with children, will do best with a puppy who scores normal in three to four tests. A puppy who scores dominant or independent in three to four tests will need firm training, and in many breeds could become overly bold and aggressive. A puppy who scores submissive in three to four tests will need very gentle, careful training, and this can sometimes be as difficult as training a dominant or independent puppy; in many breeds, an overly submissive, timid, or nervous puppy can mature into a skittish adult or a "fear biter" who snaps when something frightens him. A puppy who is a real mixed bag of labels will probably be unpredictable in his response to training and unusual situations; since it may be hard to "read" this puppy, he's not a good choice for a novice.

Caution: Keep in mind the breed you're testing. In a dominant or independent breed, the responses of most of the puppies may be dominant or independent. In a gentle and submissive breed, the responses of most of the puppies may be submissive. Indeed, such a uniform response is an indication that most of the litter will turn out to have the dominance, independence, or submissiveness typical for their breed. This type of uniform litter is a tribute to the breeder and makes the buyer's choice much easier.

I once went straight from temperament-testing a litter of Shetland Sheepdogs who sniffed hesitantly at me, to a litter of Norwegian Elkhounds who only let me pet them for a moment before striding across the lawn to do their own things. They were two nice, typical, uniform litters. But make sure the uniformity is the right one for the breed: If the Shelties had been independent and the Elkhounds submissive, something would have been wrong!

Final recommendation: Even in dominant breeds, submissive breeds, and independent breeds, most families would still do best with the puppy whose responses lean toward the normal.

After you've tested each puppy's temperament, give him a brief physical exam.

His *skin* should not have red splotches or hairless patches, and he should not be scratching furiously.

His *coat* will be a "puppy" coat, shorter and thinner than an adult's. In a short- or medium-haired breed whose standard warns against a long coat, long tufts of hair or fur around his ears will tip off a future long coat.

His *color* may be different than an adult's; many breeds take months to develop their correct colors. Check your breed profile to see if your breed is one in which a puppy starts out lighter or darker than his adult color.

His *eyes* should not be runny. They may still be puppy-blue, or they may be changing toward his adult color.

Ears that stand up by themselves—prick ears—may not be up yet. Some breeds have them up at six weeks, while others do not have them up until twelve weeks (especially in the larger breeds). When they do come up, they may come up quarter-mast or half-mast for a while until they finally straighten all the way up. Also, ears sometimes go up and down during teething. But beware of a puppy of four months or older with heavy ears that are fully hanging—they may not come up at all, or they may need extensive taping and bandaging to bring them erect.

The *teeth* of most breeds should meet in a scissors bite, but some breeds prefer or allow a level (even) bite. In a seven- to twelve-week-old puppy, the bite may already be scissors or level, or it may be slightly overshot, where the upper front teeth jut out over the bottom front teeth by about the distance of a wooden match head. Slightly overshot usually corrects itself into a normal bite as the puppy grows up.

Very far overshot, where the upper front teeth jut way out over the bottom front teeth, will not correct itself. And any degree of undershot, where the bottom front teeth jut out beyond the upper front teeth, will not correct itself. Poor-quality puppies very often have severe overshot or undershot bites, so checking a puppy's bite is always a good test of a litter.

A *dewclaw* is a fifth claw on the ankle. Most puppies have them on their front legs, and that's fine. But dewclaws on the hind legs are not fine; the theory is that when a dog scratches himself with his hind leg, the dewclaw could tear off. So unless you're looking at one of the few breeds that *requires* dewclaws on the hind legs, make sure the breeder has had any hind dewclaws removed by the vet. This is another sign of a good breeder.

Tails should have been docked at a few days old. Tails that are supposed to be furry and curled over the back will seem thin when compared to an adult's, but the fur fills in later.

Testicles should be down by twelve weeks, although they may go back up when the puppy gets excited. If both have come down at least once, they'll eventually come down to stay. If you really don't want to chance having a monorchid or cryptorchid dog, don't buy the puppy, because even a written guarantee that the testicles will come down cannot bring them down!

After you've temperament-tested and physically examined each puppy, tell the breeder that you and your family would like a few minutes alone to talk things over and make a decision.

11

Evaluating an Older Dog

If you're looking at an older dog (over four months), you'll test him and examine him a little differently. Here's how:

1. Ask the breeder to hold the dog on the leash while you walk out of sight. After a few minutes, come out of hiding and return, looking directly at the breeder, paying no outward attention to the dog (but watching him out of the corner of your eye). Say hello to the breeder, chat for a minute, then leave, all the time paying absolutely no attention to the dog.

    Depending on the breed, the dog may bark when he first sees you, then act friendly, indifferent, or reserved. He may watch you carefully, but he should never try to bite or bolt to the end of the leash, and his tail should never be tucked between his legs (except in breeds like Greyhounds, whose tails are often naturally carried between their legs).

2. Go out of sight, wait a few minutes, then repeat the test, but this time direct your gaze and attention entirely to the dog. Say hello to him, tell him how beautiful he is, and pet him. He may still act friendly, indifferent, or reserved, but he should never be aggressive or fearful.

3. Ask your children to join you. Observe his reaction to their petting and playing. He should never nip at them, even in play, and he should never be skittish of them.

4. Go out of sight, wait a few minutes, then have the breeder walk the dog past your hiding place. As they pass by, shake a coffee can full of pebbles. Normal responses: The dog may be alert to the sound, he may startle a little, he may hardly look up, he may try to investigate, or he may continue on his way. He should never bolt to the end of the leash or stand trembling with his tail tucked and eyes wide. A dog who exhibits these responses could bite defen-

sively if someone (a child, for example) surprised him with a sudden noise.

5. Come out of your hiding place, pulling a rattling pull-toy or other unusual object across the path of the dog and breeder. As with the sound test, he should be either interested or indifferent but not afraid. He'll see many new things in the world, and he must be willing to accept them with equanimity.

6. Take his leash and go for a stroll around the yard or along the sidewalk. Notice how he reacts to cars, bikes, people, rustling leaves. Ideally, he will be calm and controllable, but if he pulls eagerly on the leash, that's at least a good sign that he's interested in and unafraid of the world around him; the pulling can be controlled with obedience training. Use your common sense, though. If his strength or excitability make you uncomfortable, if you're uncertain of your ability to hold onto him, he may be too much dog for you. You'll be living with this dog, walking him, training him, for a long time. Don't get a dog who makes you uncomfortable or uncertain.

7. Ask the breeder if you and your family can sit alone with the dog in the yard or house. Keep him on the leash. Pet him, talk to him, try to get a feel for his reactions, his habits, his idiosyncrasies. Depending on his breed, personality, and age, he may not be very outgoing with you yet, but that will change with time, so don't let his reserve bother you. After all, you're a complete stranger to him. The important question is: Do you all like him?

8. Take this quiet time to examine him physically. His skin should be free of red splotches and hairless patches, and he should not be scratching furiously. If he's an adolescent, his size, coat, and color may not be fully developed yet, but his ears, eye color, nose color, bite, and tail should be correct for his breed unless he's being sold as a pet precisely because one of these is faulty. Both testicles should be down in a male, unless he's been neutered. He should not be limping or coughing.

9. Double-check with the breeder the questions you asked on the telephone about housebreaking, training, barking. See if the breeder's assurances agree with your own observations; you're looking for a breeder who represents his dogs honestly. Ask if there is a money-back guarantee if the dog doesn't get along with your other pets within a reasonable time.

10. Tell the breeder that you and your family would like a few minutes alone to talk things over and make a decision.

Saying Yes...Saying No

Yes or no?

If you decide that a puppy or dog is perfect for your wants, needs, personality, and lifestyle, congratulations! You've made the most informed decision a prospective dog buyer could ever make, and you're to be congratulated for your time and effort.

Be sure the breeder fills out and signs the blue registration application (or white registration certificate) and the written guarantee. Be sure you read the sales contract (and possibly spay contract) before signing them. Be sure you have a copy of the pedigree and health certificate. Be sure you've seen the OFA, CERF, or VWD certificates of the sire and dam. Be sure you get a receipt for your money.

You've just bought yourself a fine purebred dog!

If you decide that none of the puppies or dogs are right for your wants, needs, personality, and lifestyle, say politely: "These are really nice puppies, but we don't see exactly what we're looking for right now. Thank you for showing them to us."

I know you're disappointed, but you'd be *more* disappointed if you chose the wrong puppy. Reflect on everything that this valuable first experience has taught you about the breed, about temperament testing, and about evaluating. Call the next breeder on your list. Take your time. The search will be worth it.

APPENDIX

National Clubs or Contacts

THE SPORTING GROUP

American Brittany Club: Velma Tiedeman, 2036 N. 48th Ave., Omaha, NE 68104

American Pointer Club: Joyce Skrob, Rt. 1 Box 76, Morrison, FL 32668

German Shorthaired Pointer Club of America: Geraldine Irwin, 1101 W. Quincy, Englewood, CO 80110

German Wirehaired Pointer Club of America: Carol Stuart, RD 2, Box 182, Altoona, PA 16601

American Chesapeake Club: 25720 W. Lehmann Blvd., Lake Villa, IL 60046

Curly-Coated Retriever Club of America: Sue Tokolics, 303 S. Concord Rd., West Chester, PA 19382

Flat-Coated Retriever Society of America: 16 N. Earlton Rd. Ext., Havre de Grace, MD 21078

Golden Retriever Club of America: Linda Atwell, 6516 Esther N.E., Albuquerque, NM 87109

Labrador Retriever Club: Dr. Bernard Ziessow, 32695 Redfern, Franklin, MI 48025

English Setter Association of America: Dawn Ronyak, 114 S. Burlington Oval Dr., Chardon, OH 44024

Gordon Setter Club of America: Jane Matteson, 632 W. El Morado Ct., Ontario, CA 91762

Irish Setter Club of America: Sam McDonald, 4700 Trail Lake Dr., Ft. Worth, TX 76133 (817) 921-0910

American Water Spaniel Club of America: Tom Olson, 434 3rd Ave. N., Milick, MN 56353

Clumber Spaniel Club of America: Edythe Donovan, 241 Monterey Ave., Pelham, NY 10803

American Cocker Spaniel Club: Margaret Ciezkowski, 12 Wood Lane South, Woodmere, NY 11598

English Cocker Spaniel Club of America: Kate Romanski, P.O. Box 252, Hales Corners, WI 53130

English Springer Spaniel Field Trial Association:

Frances Nelson, 3146 Arthur St., Minneapolis, MN 55418

Field Spaniel Society of America: Carol Durkin, 223 Wedgewood Dr., Toms River, NJ 08753 (201) 341-1795

Irish Water Spaniel Club of America: Susan Tapp, 434 Webster Ave., Washington Township, NJ 07675

Sussex Spaniel Club of America: Linda Legare, 4488 280th St., E. Randolph, MN 55065

Welsh Springer Club of America: M. Pencak, Old Forestburg Rd., Sparrowbush, NY 12780 (914) 856-4533

Vizsla Club of America: Jan Bouman, 14808 Williamsburg Curve, Burnsville, MN 55337 (612) 435-7806

Weimaraner Club of America: Dorothy Derr, P.O. Box 6020, Cheyenne, WY 82003 (307) 634-8814

Wirehaired Pointing Griffon Club of America: 11739 SW Beaverton Hwy 201, Beaverton, OR 97005 (503) 629-5707

THE HOUND GROUP

Afghan Hound Club of America: Norma Cozzoni, 2408A Rt. 31, Oswego, IL 60543

Basenji Club of America: Ms. Marti Reed, 13948 W. 73rd Place, Arvada, CO 80005

Basset Hound Club of America: Andrea Field, 6060 Oak Hill Lane, Centerville, OH 45459

National Beagle Club: Joseph Wiley, Jr., River Rd., Bedminster, NJ 07921

American Black and Tan Coonhound Club: Geraldine Kline, 18436 N. 42nd St., Phoenix, AZ 85032

American Bloodhound Club: Ed Kilby, Rt. 1, Box 72 Biki, Daytona Beach, FL 32124

Borzoi Club of America: Alice Reese, 5200 Sandpiper Lane, Birmingham, AL 35244

Dachshund Club of America: Julia Goulder, 7540 Silvercrest Way, Scottsdale, AZ 85253

American Foxhound Club: Mrs. Jack Heck, 1221 Oakwood Ave., Dayton, OH 45419

English Foxhound: No club.

Greyhound Club of America: Dr. Elsie Neustadt, 334 King St., Box 2075, Hanover, MA 02339

REGAP (Retired Greyhounds As Pets): Box 2137, Short Beach, CT 06405 (203) 467-7407

Racers Recycled: Janet Huey, Box 270107, Houston, TX 77277-0107

Harrier: No national club; contact Seaview Harriers, Ventura, CA (805) 642-8758

Ibizan Hound Club of the United States: Karen Mele, Rt. 1 Box 349 Springfork Rd., Granville, TN 38564

Irish Wolfhound Club of America: 3690 Woods Rd. East, Port Orchard, WA 98366

Norwegian Elkhound Association of America: Mary Anne Ammann, 3101 W. Clinton St., Phoenix, AZ 85029

Otterhound Club of America: Cindy Chrisos, 16306 Collins Rd., Woodstock, IL 60098 (815) 337-0519

Petit Basset Griffon Vendeen: Barbara Wicklund, 1737 Rt. 206, Skillman, NJ 08558

Pharaoh Hound Club of America: Rita Sacks, Rt. 209, Box 285, Stone Ridge, NY 12484 (914) 687-9200

Rhodesian Ridgeback Club of the United States: Natalie Carlton, P.O. Box 236, Cortaro, AZ 85652

Saluki Club of America: Marjorie Seitz, 18 Morris Ave., Box 236, Cold Spring, NY 10516

Scottish Deerhound Club of America: Catherine Doyle, 6000 Northland Ave. NE, Albuquerque, NM 87109

American Whippet Club: Phoebe Booth, 2603 Bell Oak Rd., Box 237, Williamston, MI 48895

THE WORKING GROUP

Akita Club of America: Pat Mucher, 6483 Rt. 20A West, Perry, NY 14530

Alaskan Malamute Club of America: Cap Schneider, 21 Unneberg Ave., Succasunna, NJ 07876

Bernese Mountain Dog Club of America: Becky Wolfert, 115 Primrose Place, Lima, OH 45805

American Boxer Club: Cathie Hazelwood, 55 Polk Ave., E. Northport, NY 11731

American Bullmastiff Association: Martha Robins, 190 Chelsea Ct., Port Charlotte, FL 33952

Doberman Pinscher Club of America: Karen Plourde, P.O. Box 10016, Torrance, CA 90505

Giant Schnauzer Club of America: M. Galuszka, 324 Oakwood Ave., W. Hartford, CT 06110 (203) 233-2286

Great Dane Club of America: Marie Flint, 442 Country View Lane, Garland, TX 75043

Great Pyrenees Club of America: Judy Brown, 3360 Jackson Dr., Jackson, WI 53037 (414) 677-2894

Komondor Club of America: Cheri Bahrke, 12995 W. Chenango Ave., Morrison, CO 80465

American Kuvasz Association: Patricia Cole, 6419 Whaleyville Blvd., Suffolk, VA 23438 (804) 986-3061

Mastiff Club of America: Jody Greene, 309 Carter St., New Canaan, CT 06840 (203) 966-4253

Newfoundland Club of America: Joan Bendure, 6765 W. Platz Rd., Fairview, PA 16415

Portuguese Water Dog Club of America: B. Rafferty, 13765 New Discovery Rd., Colorado Springs, CO 80908 (719) 495-4264

American Rottweiler Club: Doris Baldwin, P.O. Box 23741, Pleasant Hill, CA 94523

Saint Bernard Club of America: Charlotte Miller, 1499 W. 83rd St., Hialeah, FL 33014 (305) 821-4694

Samoyed Club of America: Kathy Carr, 6345 Hart Rd., Rockford, IL 61111

Siberian Husky Club of America: Carol Nash, 54 Newton Rd., Plaistow, NH 03865

Standard Schnauzer Club of America: Barbara Hendrix, 105 Sheffield Rd., Cincinnati, OH 45240 (513) 771-4295

THE TERRIER GROUP

Airedale Terrier Club of America: Phyllis Madaus, 205 Satsuma Dr., Sanford, FL 32771

American Staffordshire Terrier Club of America: Jane Rebello, 22740 Vial Circle, Moreno Valley, CA 92387

Australian Terrier Club of America: Carol Sazama, 3 Pin Oak Trail, Medford, NJ 08055

Bedlington Terrier Club of America: A. Norman Rappaport, 25 Delaware Trail, Denville, NJ 07834 (201) 625-2667

Border Terrier Club of America: Jean Clark, RD 3, Box 572, Weare, NH 03281

Bull Terrier Club of America: Betty Desmond, RD 2, Box 315, Claysville, PA 15323

Cairn Terrier Club of America: Mrs. T. E. Lain, Box 18, Cleburne, TX 76033 (817) 645-7877

Dandie Dinmont Terrier Club of America: Mrs. Brewer, 1016 Mars Dr., Colorado Springs, CO 80906 (719) 632-4592

Irish Terrier Club of America: Peggy Gill, 166 Judy Farm Rd. Box 22, Carlisle, MA 01741 (508) 369-3006

United States Kerry Blue Terrier Club: JoAnn Custer, 602 W. Fernwood Dr., Toronto, OH 43964

United States Lakeland Terrier Club: Phyllis Belden, P.O. Box 87, Norge, VA 23127 (804) 564-8569

American Manchester Terrier Club: Beverly Longuil, 11165 S.W. 69th Terrace, Miami, FL 33173

American Miniature Schnauzer Club: Jane Gilbert, 5 Salt Meadow Way, Marshfield, MA 02050

Norfolk and Norwich Terrier Club: Edward Resovsky, 12 W. Southampton Ave., Philadelphia, PA 19118

Scottish Terrier Club of America: Bill Martin, 1041 E. Main, Flushing, MI 48433

American Sealyham Terrier Club: Marcia Jewell, Lindley Rd. RD 1, Box 90, Canonsburg, PA 15317 (412) 745-7070

Skye Terrier Club of America: Ann Boucher, 16329 Temple Terrace, Minnetonka, MN 55345

American Fox Terrier Club (Smooth and Wire): William Potter, 721 Plantmore, St. Louis, MO 63135

Soft-Coated Wheaten Terrier Club of America: 508 N. Ohio St., Hobart, IN 46342

Staffordshire Bull Terrier Club: Robert Logan, 3412 W. Clark Ave., Burbank, CA 91505

Welsh Terrier Club of America: Helen Chamides, 698 Ridge Rd., Highland Park, IL 60035 (312) 831-0313

West Highland White Terrier Club of America: Susan Napady, 5870 U.S. 6, Portage, IN 46368

THE TOY GROUP

Affenpinscher Club of America: Mrs. Beskie, Suite 400 Lincoln First Tower, Rochester, NY 14643

American Brussels Griffon Association: 5921 159th Lane N.W., Anoka, MN 55303 (713) 783-8887

Chihuahua Club of America: Janet Mohr, 4488 Schneider Rd., Ann Arbor, MI 48103

English Toy Spaniel Club of America: Susan Jackson, 21124 Shore Acres Rd., Cassopolis, MI 49031

Italian Greyhound Club of America: Bonnie Dennison, 6904 Vancouver Rd., Springfield, VA 22152

Japanese Chin Club of America: Charla Vaughan, 255 Bolivar 149, Salinas, CA 93906

American Maltese Association: Bonnie Shumard, 7546 El Escorial Way, Buena Park, CA 90620

American Manchester Terrier Club: Beverly Longuil, 11165 S.W. 69th Terrace, Miami, FL 33173

Miniature Pinscher Club of America: Nancy Mathieu, 5500 Bentley Ave., Las Vegas, NV 89122

Papillon Club of America: Roseann Fucillo, 2600 Kennedy Blvd., Jersey City, NJ 07306

Pekingese Club of America: Miss Hetty Oringer, 3 Carolyn Terrace, Southboro, MA 01772

American Pomeranian Club: Audrey Roberts, 7559 Pasito Ave., Rancho Cucamonga, CA 91730

Poodle Club of America: Mrs. Richard Kiczek, S. New Boston Rd., Box 211, Francestown, NH 03043

Pug Dog Club of America: Ann White, Rt. 13 Box 45A, Lincoln, NE 68527

American Shih Tzu Club: Betty Blair, P.O. Box 431469, Miami, FL 33243

Silky Terrier Club of America: P.O. Box 1132, Alameda, Ca 94501

Yorkshire Terrier Club of America: Betty Dullinger, RFD 2, Box 104, Kezar Falls, ME 04047

THE NON-SPORTING GROUP

Bichon Frise Club of America: Bernice Richardson, Rt. 2, Gulch Lane, Twin Falls, ID 83301

Boston Terrier Club of America: Nancy Washburn, Orange Rd., New Salem, MA 01355

Bulldog Club of America: Karlene Patterson, Ranchland Rd., Rt.4, Box 229C, Roanoke, TX 76262

Chow Chow Club: Jill Stillwell, Rt. 3 Box 306-B, Dallas, NC 28034

Dalmatian Club of America: Marjorie Doane, 325 Old Mill Spring Rd., Rt. 3, Jonesborough, TN 37659

Finnish Spitz Club of America: Bette Isacoff, 147 Brough Rd., Cheshire, MA 01225 (413) 499-4173

French Bulldog Club of America: David Kruger, 816 Fairfax Dr., Gretna, LA 70056

Keeshond Club of America: Pat Yagecic, 4726B Grant Ave., Philadelphia, PA 19114 (215) 637-7731

American Lhasa Apso Club: Sally Silva, 3197 First St., Riverside, CA 92507

Poodle Club of America: Mrs. Richard Kiczek, S. New Boston Rd., Box 211, Francestown, NH 03043

Schipperke Club of America: Judith Vossler, 13800 Bellevue Dr., Minnetonka, MN 55345

Tibetan Spaniel Club of America: Herbert Rosen, 11404 Lhasa Lane, Lutherville, MD 21093

Tibetan Terrier Club of America: Marteen Nolan, 100 West U.S. Hwy 6, Valparaiso, IN 46383

THE HERDING GROUP

Australian Cattle Dog Club of America: P.O. Box 990716, Redding, CA 96099

Bearded Collie Club of America: Patty Carmejoole, 80 Union Ave., Sudbury, MA 01776 (508) 443-5700

American Belgian Malinois Club: J. Robles, 22410 S.E. Auburn, Black Diamond Rd., Auburn, WA 98002 (206) 886-1316

Belgian Sheepdog Club of America: Gloria Davis Bray, 9831 Sadler Lane, Perry Hall, MD 21128

American Belgian Tervuren Club: Mrs. P. J. Laursen, 77017 Omo Rd., Armada, MI 48005

American Bouvier des Flandres Club: Ellen Raper, 1718 Trinity Rd., Raleigh, NC 27607

Briard Club of America: Sue Erickson, P.O. Box 3373, Mankato, MN 56002

Collie Club of America (Rough): John Honig, 72 Flagg St., Worcester, MA 01602

American Smooth Collie Association: Jan Anders, 1368 Valley View Rd., Ashland, OR 97520 (503) 482-1435

German Shepherd Dog Club of America: Blanche Beisswenger, 17 W. Ivy Lane, Englewood, NJ 07631

Old English Sheepdog Club of America: 7 Skyline Dr., Burbank, CA 91501 (713) 996-7012

Puli Club of America: Carolyn Nusbickel, 8078

Goshen Rd., RD 2, Malvern, PA 19355 (215) 296-8425

American Shetland Sheepdog Association: Peter Hanson, 19862 LeMans Circle, Yorba Linda, CA 92686

Cardigan Welsh Corgi Club of America: Bonnie Scherer, Rt. 3, Box 271-F-5, Sumter, SC 29154

Pembroke Welsh Corgi Club of America: Dr. John Vahaly, 1608 Clearview Dr., Louisville, KY 40222

THE MISCELLANEOUS GROUP

Border Collie Club of America: Janet Larson, 6 Pinecrest Lane, Durham, NH 03824

Canaan Club of America: Lorraine Stephens, Box 555, Newcastle, OK 73065 (405) 387-5576

Great Lakes Canaan Club: Bryna Comsky, 565 Illinois Blvd., Hoffman Estates, IL 60194 (312) 885-8395

Cavalier King Charles Spaniel Club: Steven Shapiro, 6624 Allott Ave., Van Nuys, CA 91401 (818) 505-1135

American Chinese Crested Club: Terry Hendricks, P.O. Box 246, LaFayette, CO 80026 (303) 666-8456

Chinese Shar-pei Club of America: Bob Calltharp, 9705 E. 80th St., Raytown, MO 64138 (816) 737-1697

Greater Swiss Mountain Dog Club of America: H. Summons, RD 8, Box 203, Sinking Spring, PA 19608 (215) 678-3631

Miniature Bull Terrier Club of America: Augustine Napoli, 826 W. Roscoe, Chicago, IL 60657 (312) 477-2106

THE RARE GROUP

American Eskimo Dogs of America: Carolyn Jester, Rt. 3, Box 211-B, Stroud, OK 74079

American Hairless Terrier Association: Becky Scott, 309 3rd St., Trout, LA 71371 (318) 992-5336

Australian Shepherd Club of America: P.O. Box 921, Warwick, NY 10990

Bolognese Club of America: Dorothy Goodale, Box 1461, Montrose, CO 81402 (303) 249-6492

Cesky Terrier Club of America: Box 178, Englewood, OH 45322

National Cesky, Terrier Club, P.O. Box 217, Maximo, OH 44650

Chinook Owners Association: T. J. and Grace Anderson, P.O. Box 3282, Jackson, WY 83001 (307) 733-3182

English Shepherd Club: 1251 Stevens Ave., Arbutus, MD 21227 (301) 242-0020

Glen of Imaal Terrier Club of North America: Frank Murphy, 1217 W. 69 St., Kansas City, MO 64113

Havanese Club: Jan Fuss, 8048 Stagecoach Rd., Crossplains, WI 53528

Jack Russell Terrier Breeders Association of America: Donna Maloney, Box 353, Windham, CT 06280 (203) 423-8350

Jack Russell Terrier Club of America: P.O. Box 365, Far Hills, NJ 07931

Kyi-Leo: Harriet Linn, 1757 Landana Dr., Concord, CA 94520 (415) 685-4019

Leonberger Club of America: Marlene Stuteville, P.O. Box 15662, Seattle, WA 98115 (206) 524-7448

National Association of Louisiana Catahoula Leopard Dogs: P.O. Box 1040, Denham Springs, LA 70727 (504) 665-6082

Little Lion Dog Club of America: Sandra Lunka, 2771 Graylock, Willoughby Hills, OH 44094 (216) 951-5288

Norwegian Buhund Club of America: 6614 Vollmer, Godfrey, IL 62035 (618) 466-3777

Norwegian Lundehund Club of America: Kimberly Richey, 9 Willow St., Littleton, NH 03561 (603) 444-2455

Nova Scotia Duck Tolling Retriever: Diana Semper, 1211 N. 146th Plaza, Omaha, NE 68154 (402) 493-4411

Polish Owczarek Nizinny Sheepdog: Kaz & Betty Augustowski, 1115 Delmont Rd., Severn, MD 21144 (301) 551-6750

Shiba Ken Club: Kathy Brown, 172 Jewett St., Akron, OH 44305 (216) 434-2921

National Shiba Club of America: Christine Eicher, P.O. Box 227, Rockaway, NJ 07866

Swedish Vallhund Enthusiasts Association: Marilyn Thell, 222 Waterman Hill Rd., Greene, RI 02827 (401) 397-5003

Toy Fox Terrier Club of America: Douglas Gordon, 111 Moose Dr., Crosby, TX 77532 (713) 324-4269

PHOTOS AND CREDITS

THE SPORTING GROUP

Brittany: Roja's Malibu Jazz Matazz, CD, owned by Gail Wilder of Malibu, CA, photo by Michael Wilder. *Pointer*: Working Service Dogs for the Handicapped: Ch. Scanpoint's Barefoot Contessa, CD and Am/Can Ch. OTCh Scanpoint's Sunrise Serenade, UD, owned by Lynn Deering of Margate, FL. *German Shorthaired Pointer*: Ch. Fieldfine Cinnamon Schnapps, SH, owned by Fred Cohen and Fieldfine Kennels of Houston, TX. *German Wirehaired Pointer*: Ch. Rosewire's Captain Tudwell, owned by Roselyn Glasgow of Huntington Beach, CA. *Chesapeake Bay Retriever*: Ch. Mitsu Kuma's Pond Mist, CD, JH, WD, Pond Hollow Dowitcher, and Ch. Pond Hollow's Spindrift, CD, SH, WD, owned by Bill and Dyane Baldwin of Newport, PA. *Curly-Coated Retriever*: Mayhem Mum's The Word, CD, TD, TDI and Ch. Charwin Paisley CR. Porsche, CDX, WC TDI, owned by Kathleen Kardash, Eclypse, of San Diego, CA. *Flat-Coated Retriever*: Ch. Petersfield Karmel Lalique, owned by Pat Debree of Killingworth, CT, photo by Kurt Anderson. *Golden Retriever*: OTCh "Gaines Top Dog" Shoreland's Big Harry Deal, owned by Sue Mayborne of Roscoe, IL. *Labrador Retriever*: Cirques Emmy, owned by Abby Kagan of New York, NY, photo by Greg Goebel. *English Setter*: Am/Can/Ber Ch. Velvet's Blue Moon, CD, owned by Sarah Anne Sly, Seal Rock, of N. Quincy, MA. *Gordon Setter*: Blackjack's Dealer's Choice, owned by Gail Clark of Fort Collins, CO. *Irish Setter*: Am/Can/Ber Ch. Can OTCh Devlin's Tamber Glow, UD, Ber CDX, owned by Mary and Lee Diesem of Rexford, NY. *American Water Spaniel*: Barth's Super Star of Swan Lake, owned by John Barth of Pardeeville, WI. *Clumber Spaniel*: Ch. Sasquatch Chaplin, CD, WDX, owned by Jane and Darrell Reeves of Roseburg, OR. *American Cocker Spaniel*: Ch. Barbandale's TC Cherub, owned by Barbara Prentiss of Kirkland, WA, photo by Vaora's. *English Cocker Spaniel*: Ch. Cabin Hill I'm

Gordon, owned by Rebec and Harvey Riggs of Costa Mesa, CA. *English Springer Spaniel*: Ch. Nokola Ziggy Stardust, UD, owned by Ron and Gayle Hutchison of Reseda, CA, photo by Art Stewart. *Field Spaniel*: Ch. Jeron's Daydream Believer, owned by Jeannine Pyles and Clint Livingston of San Antonio, TX. *Irish Water Spaniel*: Ch. Co-R's Maeve O'Blu Max, CD and Ch. Co-R's Sullivan O'Blu Max, CD, owned by Ruth Roes of Santa Barbara, CA. *Sussex Spaniel*: Ch. Topjoys Sussex Mooncloud, owned by Kathy Perry and Ruth Gardner, Teekay, of Ponca City, OK. *Welsh Springer Spaniel*: Ch. Colwyn's Winchester Special, owned by Lisa and David Hubler of San Clemente, CA, photo by Ernie Lowell. *Vizsla*: Ch. Popple Dungeon Paul Revere, VC, owned by Francis and Pat Carnes, Mattapex, of Stevensville, MD. *Weimaraner*: Ch. Moonshadow's Heaven-Scent, CDX, NSD, RDX, VX, owned by Valerie Fare of of Grand Prairie, TX, photo by Sammy Fare. *Wirehaired Pointing Griffon*: Baron Fratzkes Chukar Magic, owned by John Heublein of Enumclaw, WA.

THE HOUND GROUP

Afghan Hound: Ch. Chaparral Soylent Blue Ro-Jan, CDX, FCH, TT, owned by Shelley Hennessy of Toledo, OH. *Basenji*: Mother and son: Ch. Sukari Raider of the Lost Bark and Ch. Sukari's Mindiana Jones, CD, FCH, owned by Kathleen Jones and Laura Richarz of Reseda, CA, photo by Laura Richarz. *Basset Hound*: Ch. Bowler's Greta Garbo, CD, owned by Kristen and Don Eskew, Four Oak, of Huntington, WV. *Beagle*: Mother and son: Gracebee's Ol'Fash'n Root Beer and Ch. Just-Wright Wayward Windy, owned by Grace Rychliski of Huntington Beach, CA, photo by Ernie Lowell. *Black and Tan Coonhound*: Ch. Sandstone's Full Volume, owned by Shelley and James Cafferty Jr. of Burbank, CA. *Bloodhound*: Ch. Etowah's Just Like John Henry, CD, TD, owned by Kathy and Harold Fleming of Canton, GA, photo by Valerie Markle. *Borzoi*: Ch. Songmaker's Syren, CD, FCH, owned by Jan McKenney of Colchester, CT. *Dachshund (Longhaired)*: Raisin'L All Knight Long and Taunaswald Cygnet V Bayreuth ML, owned by Mary Pyle and Irva McDougald of Lancaster, CA. *Dachshund (Smooth)*: Ch. Fleming's Turbo Diesel, owned by Stephen and Cheryl Shultz of Fountain Valley, CA, photo by Mike. *Dachshund (Wirehaired)*: Gejan's Vanna Jo Beara, owned by Jerri Smith and Jan Oswald of Cabazon, CA. *American Foxhound*: Am/Mex Ch. Brown's Mr. Mighty, owned by Christian and Rebecca Blatter of Redlands, CA. *English Foxhound*: Am/Can Ch. Crackerland Trailblazer, owned by Sandra Zornes M.D. and Terrell Templin, Our Gang, of Americus, GA. *Greyhound*: Mother and son: Huzzah Reddy Willing And Able, UD, and Ch. Huzzah Ishidot Get Up and Glow, UD, owned by Charlene Vincent of Chatsworth, CA. *Harrier*: Ch. Landlubber's Liberty, owned by Betty Burnell, Seaview, of Ventura, CA, photo by Steve Eltinge. *Ibizan Hound*: Ch. Smotare's Scarlet PJ's, FLD CH, owned by Nancy Hiles of Cincinnati, OH, and Seti, FLD CH, owned by Will and Sarah Martens. *Irish Wolfhound*: Sunstag Rysheron Breezin XU, owned by Dixie Hirsch of Silverado, CA. *Norwegian Elkhound*: Ch. Statton's Jaded Watters, UD, Can CD, owned by Stan and Cotton Silverman of Rehoboth, MA, photo by Stan Silverman. *Otterhound*: Am/Can Ch. Follyhoun First in Line, owned by Louise and Rex De Shon Jr. of St. Joseph, MO. *Petit Basset Griffon Vendeen*: Kasani Honette, owned by Valerie Link of Pleasant Hill, CA.

Pharaoh Hound: Ch. Auten's Aladdin, owned by Ron and Carol Auten of Oxford, MI. *Rhodesian Ridgeback*: Filmmaker's Starlet, owned by Gerry Gilmore and Kate Graham of Los Angeles, CA. *Saluki*: Emir of Muyamman, CD and Ch. Muyamman Rafiq Zawbaah, CD, LCM, TT, owned by Carolyn Brown of Dandridge, TN. *Scottish Deerhound*: Fairyfort's Black Narcissus, owned by Madelyn Larsen of New York, NY, photo by Caroline Brown. *Whippet*: Littermates: Am/Can/Ber/St Ch. Topnotch McFadden and Am/Can/Ber/St/Mex/Int Ch. Topnotch Buttons & Bows, TT, owned by Ellen Frenkel and Pamela Miller of White Plains, NY, photo by John Ashbey.

THE WORKING GROUP

Akita: Ch. Masumi's Lady Sunshine, ROM, owned by Judythe Dunn, Koma-Inu, of New Ipswich, NH. *Alaskan Malamute*: Five Generations: Ch. Poker Flat's Deputy Dan, WPD, WWPD, WLD, WTD, Ch. Poker Flat's Ace of Spies, CD, WPD, WWPD, WLD, WTD, Poker Flat's The Spy Who Loves Me, Ch. Poker Flat's Nahanni, CD, and Ch. Poker Flat's Cosmic Ray, CD, owned by Robin Haggard and Jim Kuehl of Champaign, IL, photo by Heritage Studios. *Bernese Mountain Dog*: U-CDX Sandusky's Brighteye Abigail, TD, TT, AM/CAN CDX, owned by Deborah Hotze of Lockport, IL. *Boxer*: Ch. Hollycrest's Stage Hand and Ch. Hollycrest's Oodles of Noodles, owned by Cheryl Colby and Leon DePriest of Riverside, IL. *Bullmastiff*: Ch. Tauralan Turkish Delight, owned by Carol Beans of Santa Ana, CA. *Doberman Pinscher*: Littermates: Easy Does It Wind Walker, UDT, Can CDX and Easy Does It Lite 'N Lacey, UDT, Can CD, owned by Bobbie Crissey of Chelmsford, MA. *Giant Schnauzer*: Ch. Skansen's Handsome Stranger, owned by Sylvia Hammarstrom of Sebastopol, CA. *Great Dane*: Ch. Waterwood's Just As I Am, CD, owned by Margaret Shappard of Fayetteville, GA, photo by Diana Alverson. *Great Pyrenees*: Ch. Summerhill's Royal Knight, owned by Lynne Gomm of San Antonio, TX. *Komondor*: Ch. Szentivani Ingo, owned by Marion Levy, Jr., Hercegvaros, of Princeton, NJ. *Kuvasz*: Littermates: Ch. Bjels-Saros Autumn Prince, B, WH, TDI and Ch. Bjel-Saros Autumn Starlight, TDI, owned by Fred and Gudrun Stein of Catawba, NC. *Mastiff*: Ch. Brite Star's Phoebe, Ch. Brite Star, and Ch. Brite Star's Titas, owned by Lance and Barbara House of Truckee, CA, photo by Vaora's. *Newfoundland*: Brassibear's Sea Cliff, UD, DD, WD, Mex PC, owned by Jeanne Dills of North Hollywood, CA. *Portuguese Water Dog*: Ch. Sun Joy's Alma, CD, TT, Ch. Neocles Duke O'Sun Joy White Cap, CD, TT, and Ch. Sun Joy's Escuro Beleza, CD, TT, owned by Alf, Beverly, and Carrie Jorgensen of Old Saybrook, CT. *Rottweiler*: Am/Can Ch. Von Worley's Alexander Am/Can CD, owned by Dawn and Jim Worley of Newark, OH. *Saint Bernard (Shorthaired)*: Ch. Beau Mar's Maggie V Dreamer, Am/Can CD, owned by Barbara and Gene Thomas of Hollister, CA. *Samoyed*: Am/Can/Int/PR/Venz/Col/SoAm/Dom/LasAm Ch. Aladdin's Solitary Man, TT, owned by Sandra and Bruce Krupski, Riverview Acres, of Rexford, NY. *Siberian Husky*: Ch. Silistra's Thief of Hearts, Am/Can CD, Mex PC, SD, owned by Gene and Karen Stinson of Elk Grove, CA, photo by Ernie Lowell. *Standard Schnauzer*: Ch. Sterling Rikki Lynn Hansen, CDX and Yamadas Robyn Lynn Hansen, CDX, owned by Richard and Judy Hansen of Apple Valley, CA.

THE TERRIER GROUP

Airedale Terrier: Int Ch. Stone Ridge Bengal Bravo, owned by Janet Johnson Framke of Wadsworth, IL. *American Staffordshire Terrier*: Ch. Rainbo Apache Brave, CD, TT, ROH and Ch. Rainbo Blue Holly, owned by Lauraine Rodgers of West Glocester, RI, photo by Jim Johnson. *Australian Terrier*: U-CD Evanz Quokka, CDX, TT, Can CD, Auz a Galah Girl, Am/Can/Mex/Int Ch. U-CD Evanz Rachel Rachel, CDX, TT, Can/Ber CD, Am/Can/Mex Ch. U-CD Evanz Butch Cassidy, CD, TT, and Evanz Intrepid Sprite, owned by Betty Jean-Roseum Harper of Ann Arbor, MI, photo by Harper and William Hart. *Bedlington Terrier*: Carillon Tyler Blue, CD, CG, owned by Donna Hurley and Lucy Heyman of Houston, TX. *Border Terrier*: Ch. Luvemur's First Edition, UD, CG, TT, owned by John and Laurale Stern of Manitowoc, WI. *Bull Terrier*: Shavin-Kingsmere Sheez It, owned by Roland and Patricia Edwards, Linda Lethin, and R. Bollong of Glendale, CA. *Cairn Terrier*: Ch. Cairmar Cowardly Lion, owned by Betty Marcum, Anna Lee Rucker, and Brenda Carroll of Alvarado, TX, photo by Pegini. *Dandie Dinmont Terrier*: Ch. Ephan Clydes Kid Colonel, owned by Patti Perkins of Ratliff City, OK. *Irish Terrier*: Kincora's Mass Maineac, owned by Susan Young and Gloria Berube of Boxford, MA. *Kerry Blue Terrier*: Am/Can Ch. Lisiji Christmas Hope, owned by Lisa and C. J. Favre and Juanita Traylor, of Cumming, GA. *Lakeland Terrier*: Ch. Kilfel Pointe Of Vu, owned by Patricia Peters of Honey Brook, PA. *Manchester Terrier (Standard)*: Am/Can Ch. Tarton of Canty, Am/Can CD, Int. Ch. Canty's Cricket, CDX, TT, and Ch. Canty's Rythum Master, owned by Dwight and Janet Varner of Morrison, IL. *Miniature Schnauzer*: Ch. Aljamar Hot Ice, CD, owned by Janice Rue and Marilyn Laschinski, Suelen-Aljamar, of Wilmette, IL, photo by Graham. *Norfolk Terrier*: Ch. Tylwyth Just Chelsea, Am/Can UD, CG, owned by Mary Fine of Storrs, CT, photo by Ashbey. *Norwich Terrier*: Ch. Camio's Keepsake, owned by Catherine Rogers of Florence, KY. *Scottish Terrier*: Ch. Stonecroft's Highland Ransom, CDX, owned by Phyllis Selby Dabbs of Bakersfield, CA. *Sealyham Terrier*: Ch. Rinklestone Ribbonetta, owned by M. Thelma Miller of Decatur, IL. *Skye Terrier*: Ch. Druidmoor Sweeney Todd, CD, owned by Charles Brown Jr. of Beverly Hills, CA, photo by Ludwig. *Smooth Fox Terrier*: Ch. Victorian Fox Hunter, UDT, owned by Stephen, Susan, & Kirstie Lytwynec of Highland, CA. *Soft-Coated Wheaten Terrier*: Am/Mex Ch. Gleangay Hi-Jinks, CDX, TT, Mex PCE, and Am/Mex/Int Ch. Dounam's Blarney-Stone, CD, Mex PC, owned by Douglas and Naomi Stewart of Whittier, CA. *Staffordshire Bull Terrier*: Ch. Yankeestaff Fireflash, owned by Mr. and Mrs. D. C. Judd, Firestaff Staffordshire Bull Terriers, of Fort Lauderdale, FL, photo by Anthony Lopez. *Welsh Terrier*: Brynhir Shelby and Am/Can Ch. Lichen Run's Laddie, owned by Bonnie Ross of Monroeville, PA. *West Highland White Terrier*: Am/Can Ch. Barley O'The Ridge, owned by Marjadele Schiele of Capron, IL. *Wire Fox Terrier*: Ch. Talludo Minstrel of Purston, owned by Janice Rue and Marilyn Laschinski, Suelen-Aljamar, of Wilmette, IL.

THE TOY GROUP

Affenpinscher: Am/Mex/Int Ch. Balu's Schwarz Wipfel Schatzen and Am/Mex/Int Ch. Balu's Schwarz Mingeri, owned by Lucille Meystedt of Rusk, TX. *Brussels Griffon (Rough)*: Jeni, owned by Darryl

Vice of Palm Springs, CA. *Brussels Griffon (Smooth)*: Cobblestone Naughty Nettie, owned by Steven and Dori West and Cobblestone Kennels of Bloomington, IL, photo by Dori West. *Chihuahua (Longhaired)*: Laguna Mucho Man of Jobarbs, owned by Susan Sine of South Laguna, CA, photo by Mike. *Chihuahua (Smooth)*: Royal Acres Tina Marie and Royal Acres Cisco Kid, owned by Joyce Flint of Seabeck, WA. *English Toy Spaniel*: Ch. Kis'n Knoble Sir Walter, owned by John and Sue Kisielewski of Monroe, VA. *Italian Greyhound*: Ch. Littleluv's Superman, CD, owned by Kathryn Holmes of West Los Angeles, CA, photo by Joan Ludwig. *Japanese Chin*: Ch. Inuhaus Fujiyama Fireball, owned by Jari Bobillot, Ur-Chin, of Portland, OR. *Maltese*: Ch. Cedarwood Sunshine Zina, owned by Marti Klabunde of Honey Creek, IA, photo by Don Petrulis. *Manchester Terrier (Toy)*: Ch. Golden Scoops G-Tia, owned by Dwight and Janet Varner, Canty, of Morrison, IL. *Miniature Pinscher*: Ch. Mahan's Master Spock, CD, owned by Barbara Mahan of Yucca Valley, CA. *Papillon*: U-CD Victory's Special Indeed, Victory's Take A Chance, TT, Ch. Victory's Second Chance, and Ch. Victory's Classic Hunter, owned by Vicki Sharinay of Los Angeles, CA, photo by Missy Yuhl. *Pekingese*: Shadow Hills Kiss of Kismet, owned by Bonnie Eriksson of Topanga, CA, photo by Mike. *Pomeranian*: Hupkas Mr. Giggilo, owned by Marian Coy, Pomaron, of Princeton, TX. *Poodle (Toy)*: Bobbil's Red Clay O'Sweeney, owned by Bobbie Wright of Munday, TX. *Pug*: Lazy G Kola, owned by Richard and Frances Weaver of Colton, CA. *Shih Tzu*: Conleys' Sweet September, owned by Sharon Conley of Mission Viejo, CA, photo by Mark Conley. *Silky Terrier*: Ch. Lyneloor's Salty Sailor, UD, owned by Mimi Elaine Lorie of San Diego, CA, photo by Joni. *Yorkshire Terrier*: Ch. Amystis Jebo's Justa Kid, owned by Becky Stone of Montgomery, TX, photo by Dick Cauter.

THE NON-SPORTING GROUP

Bichon Frise: Am/Mex/Int Ch. C & D's Count Kristopher, owned by Dolores Wolske of Elwood, IL. *Boston Terrier*: Ch. Bonnie's Jet Pilot and Ch. Bonnie's Touch of Magic, owned by Rita Otteson and Mae Wiger of Anaheim, CA. *Bulldog*: Brynne's Sir Winston, owned by Vicki and Tony Ruiz of Laguna Niguel, CA. *Chow Chow*: Ch. Don-Lee Chowtime, owned by Desmond Murphy and Susie Donnelly of Monroe, NY. *Dalmatian*: Am/Bah Ch. Ravenwood Yankee Clipper, CDX, Bah CD, owned by Kathy and Lee McCoubrey of Elbert, CO. *Finnish Spitz*: Am/Mex/Int Ch. Arkle's Firefox of Subira, CD, owned by Richard and Dawn Woods of Gardena, CA, and Mex/Can Ch. Audacious Arkle CD, owned by Michele Sevryn and Judi Burll, Inua, of Redding, CA, photo by Judith Lee Smith. *French Bulldog*: Ch. Smokey Valley's Bullwinkle, CD, owned by Richard and Michelle Shannon of Toledo, WA. *Keeshond*: Winsome's Here Comes Trouble, CD, and Rakker's Most Excellent Promise, CD, owned by Connie Jankowski of Irvine, CA, photo by Missy Yuhl. *Lhasa Apso*: Barjea's Ebony, CD, owned by Barbara Peterson of Ramona, CA. *Poodle (Miniature)*: Genals Gena La Beauté Noire, Am/Mex UD, owned by Connie Sullivan of Mission Viejo, CA. *Poodle (Standard)*: Lido's Miss Strawberry Pic, CDX, owned by Susie Osburn of Las Vegas, NV. *Schipperke*: Ch. Jetstar's Jumpin Jehosphat, owned by Mary Ann and Terrie Johnson of Mission Viejo, CA, photo by Ernie Lowell. *Tibetan Spaniel*: Ch. Tashi Tuxedo, Ch. Tashi Tao Jako, and Ch. Tashi Tao Kato, owned by Jeanne Holsapple of New Castle, IN. *Tibetan Terrier*: Ch. Kyirong's

Cedarbar Shogun, owned by Excalibur Show Dogs, Linda Immel and James Joseph, of Green Bay, WI, photo by David Gossett.

THE HERDING GROUP

Australian Cattle Dog: Ch. Jilaroo Asti's First Edition, VQW, owned by Peg and Gary Nattress, Jilaroo, of Carrollton, GA. *Bearded Collie*: Ch. Ha'Penny Moon Shadow, owned by Robert W. Greitzer and J. Richard Schneider of Riverdale, NJ. *Belgian Malinois*: Am/Can/Dom Ch. Debalex's Rainbeau D'Or, owned by Deborah Alexander of Oak Ridge, TN, photo by Michele Perlmutter. *Belgian Sheepdog*: Ch. OTCh U-UD Qazar Charfire V Siegestor, BH, WH, HIC, TT, SchHI, HOF, owned by Elaine Havens of Orange, CA, photo by Jan Haderlie. *Belgian Tervuren*: Ch. Patana's Quest For Liberty, UD, owned by Lillian White of Bellflower, CA. *Bouvier des Flandres*: Ch. Briarlea Norstar Rose O'Luke, CD, TD, owned by Sunny and Jack Rozycki of Rice, MN. *Briard*: Ch. Mon Jovis Digby de L'Etat D'Or owned by Donavon Thompson and Mary Lopez of Pacifica, CA, photo by Fox and Cook. *Collie (Rough)*: Holmhaven Magic Flash, UD, owned by Lily Sayre of Fort Lauderdale, FL. *Collie (Smooth)*: Ch. Keepsake Once Isn't Enuf, owned by Tom, Vicky, and Heather Newcomb of Lawndale, CA. *German Shepherd Dog*: Valiantdale's Michelle, CDX, Valiantdale's Katie, UD, SchH III, Valiantdale's Icon, UDT, SchHIII, and Valiantdale's Shadow, CDX, SchH I, owned by Kathy Watson of Tulsa, OK. *Old English Sheepdog*: Ch. Love'N Stuff Ringleader, owned by Richard Esquibel, Alma Aranda, and Marilyn and Kristi Marshall of San Jose, CA. *Puli*: Ch. Bowmaker's Ted E. Bear, CD, owned by Sherry Gibson of Manchester, TN, photo by Rayleen Hendrix. *Shetland Sheepdog*: Ch. Minos Remember My Name and Ch. Minos The Escape Artist, owned by Chris Gabrielides of Rancho Cucamonga, CA. *Cardigan Welsh Corgi*: Ch. Coedwig's Buckeye, owned by Kim Shira of Kirksville, MO, photo by Chelsea Thompson. *Pembroke Welsh Corgi*: Ch. OTCh Aberdare Eliza of Taliesin, owned by Peggy McConnell of Dallas, TX, photo by Bryce Beasley.

THE MISCELLANEOUS GROUP

Border Collie: Atlastafarm's Roxy Bandit, UD, owned by Marcia Kearney of El Paso, IL, photo by Paul Meier. *Canaan Dog*: Ch. Arad of Shaar Hagai, Ch. Terramara's Bekah, CD, Ch. Terramara's Calah, and Terramara's Ezraela, owned by Terry Bagley of Alberta, Canada, photo by Steve Wusyk. *Cavalier King Charles Spaniel*: Father and daughter: Ch. Glencroft Captain Cook of Kilspindie and Kilspindie Miranda, owned by Elizabeth Spalding of Falmouth, ME. *Chinese Crested*: Littermates: Ch. Gingery's Truffles 'n Cream, CD, TT, TDI and Ch. Gingery's Maple Syrup, TT, TDI, owned by Arlene Butterklee and Victor Helu of Ronkonkoma, NY, photo by Victor Helu. *Chinese Shar-pei*: U-CD Ch. Yu-No-Who Tang's Tony, CDX, owned by Jean and Richard Hamilton of Elgin, IL. *Greater Swiss Mountain Dog*: Ch. Mex Ch. Rigi v Bothswain, CD, TT, owned by Doug and

Sandy Nyeholt of Walnut, CA. *Miniature Bull Terrier*: Graymoor Genghis, owned by Donly Chorn of Lake Villa, IL.

THE RARE BREEDS

American Eskimo Dog (Standard): Grand Ch. "PR" Sierra's Snow Shadow, owned by Charline Dunnigan of Salem, OR, photo by Carl Lindemaier. *American Hairless Terrier*: Trout Creek's Boss and Trout Creek's Jay Are, owned by Edwin Scott of Trout, LA. *Australian Shepherd*: Ch. Las Rocosa Little Wolf, CD, STD, owned by Jeanne Joy Hartnagle of Boulder, CO. *Bolognese*: Jabir's Bo-Blanc Petienne, owned by Nancy and Norman Holmes of Penrose, NC. *Cesky Terrier*: Pebbles, owned by John and Lori Moody, Blarneygem, of Goldsboro, NC. *Chinook*: Father and son: Winterset's Hassad and Benjamin's Kuska, owned by T.J. and Grace Anderson of Jackson Hole, WY. *English Shepherd*: U-CD Stonehaven's Heather, HIC, TT, owned by Pamela Burgess of Flint, MI, photo by Lee's. *Glen of Imaal Terrier*: Ch. Glen Tyr's Kelly Callahan, owned by Mary Brytowski of Worcester, MA. *Havanese*: Havana's John Henry and Havana's Abby, owned by Dorothy Goodale of Montrose, CO. *Jack Russell Terrier*: Wanderground Piranha and Wanderground Terminator, owned by Piper Woods of Albany, OH, photo by Skip Newbrough. *Kyi-Leo*: Mig-Gi Mui of Lin-Kai and Nan-Di of Lin-Kai, owned by Al and Harriet Linn of Concord, CA. *Leonberger*: Argus von Klingleberg, owned by Jim and Marlene Stuteville of Seattle, WA. *Louisiana Catahoula Leopard Dog*: Koons Flagg and Koon's Gandy Dancer, owned by Kriston Von Wise of Fort Lauderdale, FL. *Löwchen (Little Lion Dog)*: Ch. Pepperland Miss Piggy, CD, owned by Earlemarie Dingel, Lambert, of Castro Valley, CA. *Norwegian Buhund*: Ch. Leifegard's Ruby, owned by Malfrid Byshja of Norway. *Norwegian Lundehund*: Lundecock's Free & Footloos, owned by Kim and Kimberly Richey of Littleton, NH. *Nova Scotia Duck Tolling Retriever*: Ch. Sylvan's Rusty Jones, Am/Can CDX, TT, WCI, Lennoxlove's Cape Sable Swan, Colony's Lennoxlove Victoria, and Am/Can Ch. Bhalgair of Lennoxlove, Am/Can CD, TT, WCI, owned by John Hamilton and Marile Waterstraat of Macedon, NY. *Polish Owczarek Nizinny Sheepdog*: Bobnuska Z Elzbieta, owned by John Richardson of Cincinnati, OH, photo by Betty Augustowski. *Shiba*: Ch. Foxtrot Sizzlin' Sin-Sation, owned by Kathy Brown of Akron, OH. *Swedish Vallhund*: Ch. Starvon Glenby, owned by Marilyn and John Thell of Greene, RI. *Toy Fox Terrier*: National Grand Ch. "PR" Gorden's Shamrock Lad, owned by Doug and Betty Gorden of Crosby, TX.

INDEX

required, questionnaire and breed selection, 20

Experience required for ownership, questionnaire and breed selection, 18

Eye contact, making, 9

Eye infections, breeds susceptible to, 139, 143–45, 149, 152, 160

Eye lacerations, breeds susceptible to, 133, 136, 138, 143, 146, 147, 152, 157, 159

Eyelash abnormalities, 30
  breeds susceptible to, 44, 51, 52, 143, 146, 147, 152, 153, 155
  vision, 6–7

Eyelid abnormalities, 30
  breeds susceptible to, 44, 45, 51–53, 64, 67, 89, 91, 95, 97, 98, 99, 102, 143, 146, 147, 152–54, 186, 201

Eye prolapse, breeds susceptible to, 136, 139, 143, 146

Eyes, 30, 244, 247
  ailments, *see specific eye ailments, e.g.,* Cataracts; Glaucoma; Retinal (eye) abnormalities, breeds susceptible to
  vision, 6–7

Eye ulcers, breeds susceptible to, 132, 134

Fanconi, breed susceptible to, 104

Faults, 27, 29, 214, 216, 231

Fear, 7

Females. *See also* Dam
  versus males, 230
  reproduction and neutering, 230–31

Field Spaniel, 54

Finnish Spitz, 155

Flat-Coated Retriever, 43

Flea allergies, breeds susceptible to, 126, 200

Flea powders, 4
  breeds sensitive to, 62, 68, 76, 81, 83

Food, 238
  breed requiring high-protein diet, 205
  breed requiring low-protein diet, 155
  breeds that should not be fed canned, 135, 139, 144, 149, 185
  costs of, 5
  hand feeding, 9

Food allergies, breeds susceptible to, 112

Foreign breeds, 14
  not recognized by AKC, 3, 4

Foxhounds. *See* American Foxhound; English Foxhound

Fox Terrier
  Smooth, 125
  Wire, 130

Fractures, breeds susceptible to, 132, 134, 135, 137, 142

French Bulldog, 157

Frisbee competition, breeds used in, 182, 192

Front and Finish, 221

Gassiness, breeds prone to, 90, 146, 152, 157

Gastric disorders, breed susceptible to, 175

Genes, 12–13, 14, 17–18, 223–24, 236

German Shepherd Dog, 22, 23, 175

German Shorthaired Pointer, 23, 39

German titles, 222–23

German Wirehaired Pointer, 23

Giant dogs, 19, 22

Giant Schnauzer, 22, 23, 92

Glaucoma, 30
  breeds susceptible to, 37, 51, 64, 65, 103, 118, 123, 125, 130, 145, 160, 179, 180

Glen of Imaal Terrier, 197

GM-1 Storage Disease, breeds susceptible to, 100

Golden Retriever, 44

Gordon Setter, 47

Grades of puppies, 223–24

Great Dane, 23, 94
  ear cropping, 4

Greater Swiss Mountain Dog, 187

Great Pyrenees, 22, 23, 94

Greyhound, 74

Greyhound Rescue Leagues, 74

Grooming, 5, 9
  questionnaire and breed selection, 20

Guarantees from breeders, 225, 238, 248

Guard dogs and home guardians, 87, 90–93, 96–98, 101, 105, 107, 108, 112, 116, 127, 154, 162, 171, 175, 177, 179, 180, 202. *See also* Watchdog